The UL 4600 Guidebook

What to Include in an Autonomous Vehicle Safety Case

Philip Koopman, Ph.D.

Carnegie Mellon University

Thanks to Deb, Jackie, Heather & Uma

First Edition, 2022. (Version 1.0.02)

Copyright © 2022 by Philip Koopman
All rights reserved.
ISBN: 9798365303065 Trade Paperback
ISBN: 9798365303249 Hardcover

No part of this publication may be reproduced or transmitted in any form or by any means, including photocopy, scanning, recording, or any information storage and retrieval system, without permission in writing from the copyright holder. This book has neither been approved nor endorsed by UL Standards and Engagement, which publishes the ANSI/UL 4600 standard.

INFORMATION IN THIS BOOK IS PROVIDED *"AS IS"* AND ANY EXPRESS OR IMPLIED WARRANTIES ARE DISCLAIMED. IN NO EVENT SHALL THE AUTHOR OR PUBLISHER BE LIABLE FOR ANY DIRECT, INDIRECT, INCIDENTAL, SPECIAL, EXEMPLARY, CONSEQUENTIAL, OR OTHER DAMAGES, EVEN IF ADVISED OF THE POSSIBILITY OF SUCH DAMAGES. THE INFORMATION IN THIS BOOK IS INTENDED TO ONLY PARTIALLY SUPPORT SAFETY PRACTICES, AND MORE IS REQUIRED TO ACHIEVE ACCEPTABLE SAFETY. YOU ARE RESPONSIBLE FOR THE SAFETY OF THE SYSTEMS YOU DESIGN AND OPERATE REGARDLESS OF THE CONTENT OF THIS BOOK. THIS BOOK IS NOT A SUBSTITUTE FOR THE OFFICIALLY ISSUED STANDARD. BY OPENING THIS BOOK, THE READER AGREES TO THESE TERMS AND ACCEPTS SOLE RESPONSIBILITY FOR ANY AND ALL DIRECT AND INDIRECT USES OF ITS CONTENTS.

Contents

Preface vii
1. **Introduction** 1
   1.1. Quick tour 1
   1.2. Resources 2
2. **Overview and applicability of UL 4600** 3
   2.1. History of UL 4600 3
      2.1.1. UL 4600 timeline 3
      2.1.2. Other AV safety standards 4
   2.2. Scope and applicability of UL 4600 5
      2.2.1. UL 4600 scope 5
      2.2.2. UL 4600 applicability 6
      2.2.3. UL 4600 and the standards ecosystem 8
   2.3. Frequently asked questions and misconceptions 9
   2.4. Resources 12
3. **Requirements and prompt elements** 13
   3.1. Requirement structure 13
   3.2. Prompt elements 14
   3.3. Deviations 16
4. **Terminology** 19
   4.1. Terminology approach 19
   4.2. Key conceptual terms 20
5. **The safety case** 23
   5.1. Structure of a safety case 23
   5.2. Safety case properties 26
      5.2.1. Safety case scope and the item 26
      5.2.2. Concept of operations 27
      5.2.3. Element Out Of Context interfaces 27
      5.2.4. Types of claims 28
   5.3. Safety case tool support 29
   5.4. Argument sufficiency 31
      5.4.1. Inapplicable prompt elements 31
      5.4.2. Assumptions 32
      5.4.3. Difficult-to-reproduce aspects 33
      5.4.4. Accepted risks 34
      5.4.5. Confirmation bias and defeaters 35
      5.4.6. Non-deductive arguments and unknowns 36
   5.5. Safety culture 37
   5.6. Resources 38
6. **Hazards and risks** 39
   6.1. Fault model 40
   6.2. Hazards 41

    6.3. Risk evaluation 43
    6.4. Risk mitigation... 44
        6.4.1. Criticality levels and risk mitigation 44
        6.4.2. Going beyond risk-based mitigation 45
        6.4.3. Acceptable item-level risk 46
    6.5. Resources 46

7. Interaction with people and road users 49
    7.1. Communication functions 49
    7.2. Interacting with people and animals 50
        7.2.1. Occupant interactions 50
        7.2.2. Passenger-settable parameters 51
        7.2.3. Demographic profile 52
        7.2.4. Safety contribution by people 52
        7.2.5. Mode changes 53
    7.3. Interacting with other vehicles 53
        7.3.1. Hazards involving other vehicles 54
        7.3.2. Rule breaking 54
    7.4. Resources 55

8. Autonomy functions and support 57
    8.1. Autonomy architecture description 57
    8.2. Operational Design Domain 58
    8.3. Sensing 60
        8.3.1. Individual sensors 61
        8.3.2. Sensor fusion 61
    8.4. Perception 62
    8.5. Machine learning 63
    8.6. Planning 64
    8.7. Prediction 64
    8.8. Trajectory, motion control, actuation, and status 65
    8.9. Timing 66
    8.10. Resources 67

9. Software & system engineering process 69
    9.1. Software and system engineering key points 69
    9.2. Resources 70

10. Dependability 71
    10.1. Fault detection and mitigation 71
        10.1.1. Fault detection coverage & diagnosis 71
        10.1.2. Reintegration 72
    10.2. Redundancy management 73
    10.3. Operational modes 74
        10.3.1. Identifying modes 75
        10.3.2. Degraded operations 75
    10.4. Item robustness 78

10.5. Incident response	79
10.6. System timing	81
10.7. Cybersecurity	81
10.8. Resources	82

11. Data and networking ... 83
- 11.1. Why data safety matters ... 83
- 11.2. Data transmission ... 84
  - 11.2.1. Data transfer properties ... 84
  - 11.2.2. Remote operation ... 85
- 11.3. Data storage ... 86
- 11.4. Infrastructure (data and otherwise) ... 87
- 11.5. Resources ... 88

12. Verification, validation, and test ... 89
- 12.1. V&V methods ... 89
- 12.2. V&V coverage ... 90
- 12.3. Testing ... 90
- 12.4. Run-time monitoring ... 92
- 12.5. Safety case updates ... 93
- 12.6. Resources ... 93

13. Tools, COTS, and legacy qualification ... 95
- 13.1. Tools ... 95
- 13.2. Simulations ... 96
- 13.3. COTS and legacy components ... 97
- 13.4. Resources ... 99

14. Lifecycle concerns ... 101
- 14.1. Requirements and design validation ... 101
- 14.2. Build, design release, and manufacturing ... 102
- 14.3. Supply chain ... 103
- 14.4. Field modifications and updates ... 103
- 14.5. Operation ... 104
- 14.6. Retirement and disposal ... 105
- 14.7. Resources ... 106

15. Maintenance ... 107
- 15.1. Maintenance and inspection requirements ... 107
- 15.2. Prompting maintenance and inspection activities ... 108
- 15.3. Maintenance and inspection faults ... 109
- 15.4. Non-operational safety ... 109
  - 15.4.1. Inactive vehicle ... 110
  - 15.4.2. Equipment degradation ... 110
- 15.5. Resources ... 111

16. Safety Performance Indicators ... 113
- 16.1. Threshold values ... 113

16.2. Dangerous behaviors	114
16.3. Surprises	115
16.4. SPI collection and feedback	116
16.5. A proposal for tying SPIs to the safety case	117
16.6. Resources	118

17. Assessment 121
    17.1. Conformance planning 121
    17.2. Self-assessment 122
    17.3. Independent assessment 123
    17.4. Continual re-assessment 125

18. Wrap-up 127
    18.1. Additional information 127
    18.2. About the author 128

Preface

Autonomous vehicles will not be viable for real-world use on public roads unless we can make them acceptably safe. Not perfectly safe to the point of zero crashes – although that is a worthy goal. Rather, acceptably safe will do for commercial deployment, with a hope of providing a substantive improvement over the current mishap rate of human drivers.

An acceptably safe outcome for autonomous vehicle (AV) deployment is not a foregone conclusion. Safety engineering does not happen all by itself, not even if super-smart engineers have the best intentions to create safe AVs. Simply being smart will not ensure all aspects of AV safety are covered any more than being smart will necessarily instill the skills and experience needed to make a safe aircraft. Achieving a safe outcome requires having strong safety engineering skills, as well as applying lessons learned across many domains in how to create safe systems.

Creating a safe system design has always required tremendous attention to detail. That in turn involves the use of specific safety engineering approaches such as hazard analysis, risk mitigation, and careful implementation of redundancy architectures. Following an industry-created safety standard helps ensure that the right approaches are used in the right way to achieve acceptable safety.

Because of the novelty of machine learning technology and lack of a human driver to display an approximation of common sense, autonomous vehicles present significantly different and dramatically more challenging issues for ensuring safety than traditional vehicles. Different companies are trying different approaches tuned for different applications and different implementation architectures. We have not yet arrived at a fixed design approach for building a safe AV. Nonetheless, such vehicles are deployed on public roads, and safety remains a pressing question.

The industry consensus at this point seems to be that safety will not be ensured by following a building code-style recipe for how to build an AV. Maybe that will happen someday, but not today. Rather, the industry has converged on the concept of a safety case as a way to argue that an AV is acceptably safe for its intended operations. The idea of a safety case for cars is not a new one. The decade-old ISO 26262 automotive functional safety standard requires a safety case.

ANSI/UL 4600 extends the safety case approach to its logical conclusion for AVs, resulting in the most comprehensive standard for autonomous vehicle safety currently available. It describes how to assess that the AV's safety case includes everything it should, to support a credible claim of acceptable safety.

This book covers the background of the standard, how the standard is structured, key terminology, and a clause-by-clause summary of the standard. It is not a detailed restatement, but rather a high-level overview. Ideally, the reader will go through this book, get the big picture, and then be ready to dive into the details of the standard itself.

To keep things concise, this is a guided tour of the standard rather than an in-depth text on system safety. The style of writing is intended to be a descriptive narrative rather than an academic text. This should make concepts more accessible to those who are not safety engineering experts. (Those looking for the humorous footnote style seen in my previous book on "how safe is safe enough" will be disappointed. This is more of a just-the-facts guided tour.)

If you are familiar with system safety concepts and functional safety, especially in the automotive industry, you might find yourself nodding along as you run down through the topics. If so, that's great, because it means you're getting the big picture in mind as preparation to dive into the standard.

If you hit a chapter on a topic you've not dealt with before, that is a great opportunity to expand your breadth in system safety before diving into the details of that part of the standard. Most chapters have a reference section with places to get started if a topic is new to you. If you are new to safety engineering for AVs in general, chapter 1 lists some getting-started resources.

Creating UL 4600 has been quite a journey. I personally wrote the proposed text (200+ pages of it) of the initial draft that kicked things off. After submitting that draft, we followed an ANSI-conformant consensus process to ensure robust engagement with stakeholders. Hundreds of comments (many hundreds) arrived from all over the world. Suggestions, and sometimes complaints, were resolved. That feedback improved clarity, added essential elements, and resolved controversy.

I have remained closely involved with the revision process for each edition via submitting change proposals, performing technical reviews, and commenting on proposed changes. But by no means am I the only one at work on this standard.

As is appropriate for an industry standard, the entire revision and approval process is public. I get one vote out of 30-40 allocated to members of the Standards Technical Panel (STP) voting committee for eventual approval of each edition of the standard. The issued standard represents the results of an accredited industry standard consensus process that reflects inputs from vehicle makers, component suppliers, regulators, consumers, assessment organizations, safety researchers, and more.

From time to time some industry politics have come into play – as they do with every standard. But a delightful thing about his process has been that the participants were overwhelmingly not there for the politics, but rather to get the job done. Many in the industry doubted that UL 4600 could be issued on our stretch-goal timeline of about a year from proposal to issued standard, but indeed that is how it turned out. It could never have been done without the common efforts, willingness to have frank discussions, and helpful contributions of so many Standards Technical Panel members and other stakeholders. Thank you so much to everyone who contributed!

Preface

If you wonder how I came to know enough about such a wide variety of topics to put into the draft proposal, chapter 18 has a brief bio. Suffice it to say that I've had a really broad range of experiences across many different industries, and seen an awful lot of stuff that is relevant to a standard like this. That includes doing hundreds of design reviews for products and components not only automotive, but also rail, industrial controls, building automation, power systems, and even a bit of work on aviation control networking safety. Much of UL 4600 falls into the bins of "don't make this mistake because that turned out badly for someone else," "this is how safety is done in other industries in addition to automotive" and "did you think of that?"

While many stakeholders made valuable contributions, there are a few special contributors I want to thank in particular. Deb Prince wrangled everyone (including me) through the standard process and supported my sometimes unconventional approaches. Jackie Erickson provided invaluable contributions to stakeholder outreach and messaging, especially with regulators and media. Heather Sakellariou did the heavy lifting on logistics, editing, comment management, and production for the standard. Uma Ferrell provided pivotal feedback on the early outline as well as lessons drawn from her extensive aviation safety experience. Thanks also to Frank Fratrik, Jason Smith, and Mahmood Tabaddor for their contributions to the drafting process. Jack Weast, Rafael Zalman, Finch Fulton, Nat Beuse, Junko Yoshida, Roger Cohen, Aaron Kane, and Chuck Weinstock also provided particularly important discussions, support, and other contributions.

Nothing is ever perfect, and everything can be improved. But fortunately, both this book and the UL 4600 standard itself can be updated with comparatively little pain. If you see something that should be fixed, please let me know via an e-mail to AVSafety@Koopman.us

Meanwhile, happy reading!

Philip Koopman
Pittsburgh, PA, November 2022.

1. Introduction

Welcome to the world of UL 4600!

This book boils a lengthy, dense, and complex standard down as much as can be done while still being a comprehensive treatment for something that is – well – lengthy, dense, and complex.

UL 4600 takes a detailed, thorough approach because its purpose is to make sure nothing important gets left out of the safety case for an autonomous vehicle (AV). AVs are incredibly complex, so there is a lot of ground to cover.

This book, in contrast, uses a more narrative approach. General themes, ideas, and considerations are described, often in an order that flows better from a narrative point of view. While most chapters correspond to the sections of UL 4600, this book's subsections do not necessarily follow the exact flow of the standard's subsections. Rather, this book follows an order better suited to telling the high-level story of each corresponding clause (chapter) in UL 4600.

Not every detail can be in this book. Rather, the main ideas are discussed in a general sense. Think of this book as a way to understand the main themes and get an orientation to what is going on, without getting bogged down in the mechanics of the standard itself. After all, if you really want the gory details, that is what the standard is for.

1.1. Quick tour

Here is a quick tour of the rest of this book:
- Chapter 2 covers the history and scope of UL 4600. Briefly, it deals with how to know that an autonomous vehicle (AV) safety case has what it needs to ensure that an AV will be acceptably safe. There is also a Frequently Asked Questions section that answers common questions and clarifies some common misconceptions regarding UL 4600.
- Chapter 3 describes the structure of UL 4600, which emphasizes the use of "prompt elements" to help remind both safety engineers and safety case assessors what should be addressed by the safety case. It is important to be oriented to this approach before diving into the standard itself.
- Chapter 4 covers key terminology and concepts. Every standard has some defined terms with nuances not necessarily easy to interpret without a little introductory guidance. Read this chapter if you want to know what UL 4600 might mean by "acceptable," "item," and "argue," among other terms.

- Chapters 5-17 cover the corresponding clauses in UL 4600, ranging from safety cases in chapter 5 to the assessment process in chapter 17.
- Chapter 18 has some pointers to additional information that might prove useful.

1.2. Resources

- UL 4600 information launch page:
  https://users.ece.cmu.edu/~koopman/ul4600/index.html
- Video tutorial on UL 4600 (23 minutes):
  - YouTube version: https://youtu.be/ZxVMX8SjPvw
  - Archive.org version: https://archive.org/details/L109-ul-4600
- Video tutorial series on AV safety to provide background
  - https://users.ece.cmu.edu/~koopman/lectures/index.html#av (includes slides, YouTube videos, and archive.org mirrors)
- A graduate-level course on embedded system and software safety taught by the author at Carnegie Mellon University, with all lecture videos freely available online: https://course.ece.cmu.edu/~ece642/
- Some historical notes on the evolution of UL 4600:
  https://www.eetimes.com/safe-autonomy-ul-4600-and-how-it-grew/

2. Overview and applicability of UL 4600

Summary: UL 4600 is primarily applicable to ensuring that the safety case for fully automated vehicle operation (including SAE Levels 3, 4, and 5) is acceptable. It is designed to work harmoniously with other relevant standards, and in particular ISO 26262 and ISO 21448.

UL 4600 has taken a somewhat different path than other safety standards. That path was carefully chosen to respond to the unique needs of the autonomous vehicle industry for a safety standard that provides flexibility, can be rapidly updated as necessary, and yet provides comprehensive guidance to ensure acceptably safe vehicles are deployed on public roads. Significant care was taken to make sure that it also plays well with other standards and practices the automotive industry depends on for success.

2.1. History of UL 4600

UL 4600 was created in response to a significant gap in the standards space regarding autonomous vehicles. While the industry was spending billions of dollars on developing the technology, it was unclear how they would ensure the safety of such systems once they got them working. One way to improve the safety outlook was to create an industry safety standard to establish a minimum acceptable practice for ensuring safety.

2.1.1. UL 4600 timeline

In late 2018 when initial work started on UL 4600, two highly relevant automotive safety standards were available. ISO 26262, with its second edition issued in late 2018, was primarily written to cover functional safety for conventional (non-autonomous) vehicles.

ISO/PAS 21448 covered Safety Of The Intended Function (SOTIF), which deals with issues relevant to driving automation features operating in the real world. ISO/PAS 21448 was scoped for advanced driver assistance system (ADAS) functions rather than fully automated driving functionality.

SAE J3016 also existed as a taxonomy and terminology standard. It defines the SAE Levels, with highly automated vehicles that are the focus of UL 4600 being assigned Levels 3, 4, and 5. However, safety engineering is not in scope for J3016.

Preliminary scoping work on UL 4600 started in summer 2018. Underwriters Laboratories (which has since changed branding and is now known as UL Standards and Engagement – ULSE) started the standardization process to ensure no conflicts with other standard development efforts. In early January 2019 there was a detailed outline. In

April 2019 there was a nearly full-size (213 pages) draft ready to receive preliminary comments from a select set of reviewers.

A fully detailed proposal was submitted as the starting point for the consensus process in May 2019. A Standards Technical Panel (STP, which is a UL standards voting committee) was constituted to consider the standard. A thorough process of two multi-day physical meetings, further discussions, and iterations of the document took place in 2019. A nearly completed version was ready for formal review in October 2019. By January 2020 there was a clean version ready for an official voting process. After some revisions and comment resolution, STP consensus was achieved, and the first edition of the standard was issued as both a UL standard and an ANSI standard on the ironic date of April 1, 2020.

ULSE is the issuing standards development organization (SDO). ULSE followed the ANSI process when developing UL 4600, and was therefore able to issue it as an official ANSI standard simultaneously. (Both UL 4600 and ANSI/UL 4600 are equivalent designations for the exact same standards document.)

There were a number of useful suggestions for further improving the standard at the time of the voting. Those that were impractical to resolve in the first edition were carried over into a revision process. Those improvements, clarifications, and other upgrades were incorporated into the 2nd edition, issued as an ANSI/UL standard on March 15, 2022. (March 15th is also known as the Ides of March. Given the date of the first edition, perhaps someone at ULSE has a dry sense of humor.)

2.1.2. Other AV safety standards

During the UL 4600 creation process, other standards organizations started issuing standards relevant to autonomous vehicles.

In January 2019, Singapore published the standard TR68: Autonomous Vehicles, which had already been in progress before work on UL 4600 became generally known. That was arguably the world's first safety standard specific to autonomous vehicles. It was an excellent step forward, but was modest in its scope, and in particular emphasized Singapore's need to have a standard for testing and approving low-speed shuttles in a specific operational area. A key leader of that effort later contributed as a member of the UL 4600 STP.

ISO/PAS 21448 for SOTIF was updated to ISO 21448, including an expansion from just driver assistance systems to also cover fully automated driving systems. That updated standard issued in June 2022. Members of the technical committee who created ISO 21448 as well as ISO 26262 also contributed to the development of UL 4600.

SAE had previously established the On-Road Automated Driving (ORAD) Committee. Among other publications they produced SAE J3016, which is a taxonomy and terminology document for automated driving systems. They also issued SAE J3018, which deals with using human safety drivers in conducting highly automated vehicle road testing.

A few automotive companies had created the Automated Vehicle Safety Consortium (AVSC), which has issued a number of guidance documents regarding automated vehicle safety. Those documents are not subject to a public consensus process, and are not accredited standards. However, some of those documents have historically informed SAE ORAD standards.

IEEE has had some involvement as well. In April 2022 IEEE issued *2846-2022 IEEE Standard for Assumptions in Safety-Related Models for Automated Driving Systems*. IEEE also has an ongoing activity in ethical design. A notable issued standard is *IEEE 7000-2021 Model Process for Addressing Ethical Concerns During System Design*. Other standards in the IEEE 7000 series both issued and in progress may be relevant, but are not specific to AV applications.

Other standards organizations have increased their emphasis on the AV industry since the start of the UL 4600 project. Those include BSI (the PAS 1880 series), ISO (including ISO TR 4804:2020 and ongoing work on ISO TS 5803), and ASAM (OpenSCENARIO and related standards).

Standards in the AV space have grown from a few efforts to what might feel like a frenzy in the space of a few years. There are other standards being developed that are not mentioned, and indeed just keeping track of all the standards activity in this space could amount to a full-time job. Nonetheless, at the time of this writing, UL 4600 remains the most comprehensive safety standard for autonomous road vehicles.

2.2. Scope and applicability of UL 4600

A narrow statement of the scope of UL 4600 is that it applies to any autonomous light road vehicle. (The 3rd edition, in the consensus process as this is written, is expected to officially extend that scope to include heavy commercial trucks that operate on public roads. This book anticipates likely changes and will remain valid for the 3rd edition of UL 4600 when it issues.)

The standard can also apply to a wide variety of autonomous ground vehicles, sometimes with only minor additions or tweaks. However, the consensus process for issuing the standard only considered the stated scope (light ground vehicles for 1st and 2nd editions, with the addition of heavy trucks to the scope for the 3rd edition). So any team using the standard beyond that scope should be sure to consider potential hazards and risks that might not have been included as prompt elements or examples.

2.2.1. UL 4600 scope

UL 4600 does not provide a prescriptive process for building an AV, nor require a specific design process to follow. Rather, it requires that a safety case be constructed, and sets forth extensive criteria to ensure that the safety case is comprehensive, valid, and sound.

Many AV standards tend to limit their scope to the Automated Driving System (ADS) as it is defined in SAE J3016. The ADS system is responsible for actually driving the vehicle. However, non-driving safety responsibilities are out of scope for J3016 and the driving automation levels (SAE Levels 1-5) that it describes. For example, post-crash safety is out of scope for the SAE J3016 level definitions.

In contrast, UL 4600 is a system-level safety standard that does not limit itself to the driving function. Human drivers are responsible for non-driving aspects of safety, such as checking cargo safety, having required conversations with law enforcement personnel, carrying out crash scene responsibilities, ensuring that required maintenance is performed, and so on. Any vehicle with an ADS that does not have a qualified human driver on board needs to have a way to address such system-level safety concerns, even if it involves remote support personnel. The scope of UL 4600 ensures that design teams do not inadvertently overlook any of these issues.

The full scope of UL 4600 can be appreciated by looking at the clause titles in the standard, the main parts of which correspond to the titles of chapters 5-17 in this book. Briefly, UL 4600 ensures that the AV's safety case captures the results of safety engineering dealing with the vehicle, its interactions with people, its design process, its dependability approach, its lifecycle support, and field engineering feedback to ensure continued safety until retirement.

Some topics within the scope of UL 4600 might be strengthened in the future with additional, more detailed content. Examples include localization, high-definition (HD) maps, and object tracking. Those functions within an AV are generally covered by UL 4600, but more specific prompt elements might be added to strengthen support for those areas in the future based on industry lessons learned. It is anticipated that the scope of UL 4600 will continue to evolve in response to industry needs and contributions of additional candidate prompt elements.

2.2.2. UL 4600 applicability

The stated scope of UL 4600 is any road vehicle that does not have continuous human supervision for safety. In practice, this means that there is no "safety driver" in the vehicle, and no remote human operator responsible for providing continuous safety supervision. This specifically includes, but is not limited to, SAE J3016 Level 4 and Level 5 vehicles (vehicles that perform the entire dynamic driving task and fallback functions automatically).

There are some additional uses of UL 4600 that might not be obvious from a brief scope statement. They include the following:
- **SAE Level 3 automation features.** During operation, an SAE Level 3 feature is fully responsible for performing the driving task safely, no differently than a Level 4 or Level 5 vehicle. While at Level 3 the human driver might be called upon to perform a Fallback function, the handoff

procedure must still be handled in a way that results in net acceptable safety. UL 4600 fully covers both the automated driving functionality (i.e., whenever the Level 3 feature is active) and the vehicle-side aspects of the handoff to a human driver. SAE Level 3 functionality clearly falls within the scope of UL 4600.

- **Teleoperated vehicles.** UL 4600 requirement 12.2.5 specifically addresses risks related to remote operator data connectivity. A significant concern is that the vehicle must maintain safety even when a teleoperation link is lost. This would apply whether the remote operator is actively driving, providing high-level guidance, or just monitoring operation of the AV remotely. The safety of any teleoperated or telesupervised vehicle is covered by UL 4600 at least with regard to the loss of the data link. The content of UL 4600 also applies to remotely operated functionality.
- **Heavy trucks.** UL 4600 2nd edition does not claim heavy trucks to be in scope. The 3rd edition increases its scope to include trucks. While some changes are being made to the 3rd edition to provide additional prompt elements and examples, no change to the overall structure of the standard was required to expand the scope, and the detailed changes are not terribly numerous. Using the 2nd edition pending publication of the 3rd edition would provide an excellent starting point for heavy truck safety cases, and could be applied successfully even in the absence of a 3rd edition with some application-specific prompt element additions (e.g., more detailed consideration of potential issues regarding hazardous loads).
- **Bike lane bots and sidewalk bots.** Smaller delivery robots and vehicles are reasonably well covered by UL 4600. To the degree they need to cross public roads and occasionally operate on public roads when sidewalks are impassable, they will need to behave as road vehicles and would be fully covered. Some specific hazards might need to be added to the safety case, such as not falling down public sidewalk staircases, and navigating sidewalk heaves that are large with respect to the bot wheelbase. But nothing in UL 4600 should prevent using it for smaller delivery bots and the like.
- **SAE Level 2+ automation features.** SAE Level 2 features control vehicle motion, with the driver providing supervision and intervening when necessary to ensure safety. While the terminology "Level 2+" is not condoned by the J3016 standard, various stakeholders use the "2+" designation to refer to highly capable automation features that approximate Level 3 or Level 4-style capabilities in series production vehicles while still holding the driver accountable for safety. Creating a UL 4600 conformant safety case for these systems can be done simply by assigning appropriate prompt element responsibilities to a human driver instead of automation. Doing so would permit checking the responsibilities assigned to human safety supervisors against the realistic capabilities of humans to do such supervision work. Specifying a

realistic limit to human capabilities is beyond the scope of UL 4600. However, the UL 4600 safety case approach would make it clear what the machine expects of a human driver so that other techniques might be used to understand how reasonable that expectation might be.

Industrial, farm, and off-road vehicles might benefit from an adaptation of UL 4600, and especially the parts that are not specific to road vehicles. Moreover, it is common for farm and other specialty vehicles to spend some of their time traversing public roads, during which times UL 4600 would fully apply. Specialty domains might require customization of prompt elements to address application-specific hazards and risk considerations. Prompt elements relevant to farm equipment might include coordination between implement and tractor, rollover hazards, and the safety of personnel obscured by vegetation.

2.2.3. UL 4600 and the standards ecosystem

The scope of UL 4600 is the entire scope of a system-level safety case for an automated driving system (minus human factors in any driving supervision). However, other standards still play a crucial role in attaining AV safety. This is particularly true of ISO 26262 and ISO 21448, but can also be true of any other safety standard that a development team finds appropriate and useful, such as MIL-STD-882.

UL 4600 explains what topics need to be covered by a safety standard. At times it gives some pretty strong hints about practices that are likely to result in unacceptable safety. It requires that a good design process be used, but does not specify any particular design process. It requires that a statement of what "safe enough" might mean be included in the safety case, but does not specify a definition for "safe enough." It suggests a myriad of potential hazards and risks that might need to be mitigated, but does not require a specific mitigation technique.

In other words, UL 4600 provides a structured framework to judge whether safety engineering has been done in a way that is likely to result in an acceptably safe AV, but does not require a specific way to get to that outcome. The how to get there part is where other standards come in.

UL 4600 specifically contemplates that ISO 26262 will be used in as many ways as are practicable for AV safety. (UL 4600 Annex A gives a partial clause mapping to ISO 26262.) The idea is that if ISO 26262 conformance is done with the requirements of UL 4600 in mind, there will be no need to do that safety work twice. ISO 26262 conformance can simultaneously satisfy many requirements in UL 4600. Anything not so satisfied needs additional work that might come from another standard, such as ISO 21448, or might come from an internal engineering practice that meets UL 4600 requirements.

From this point of view, UL 4600 ensures coverage of all relevant safety topics while permitting taking maximum advantage of processes required by other safety standards such as ISO 26262 and ISO 21448. (For government systems work, MIL-STD-882 is also a viable foundation that can be mapped to UL 4600.) Any prompt elements unsatisfied by UL 4600 after applying those standards are gaps in the standards ecosystem that the team can fill in using their best practices. Many such gaps will be at the level of system safety, at least for now.

Figure 2.1 shows a conceptual view of various AV standards and how they work together. (Not all standards discussed earlier are included – this is just a very high-level view of some key standards.)

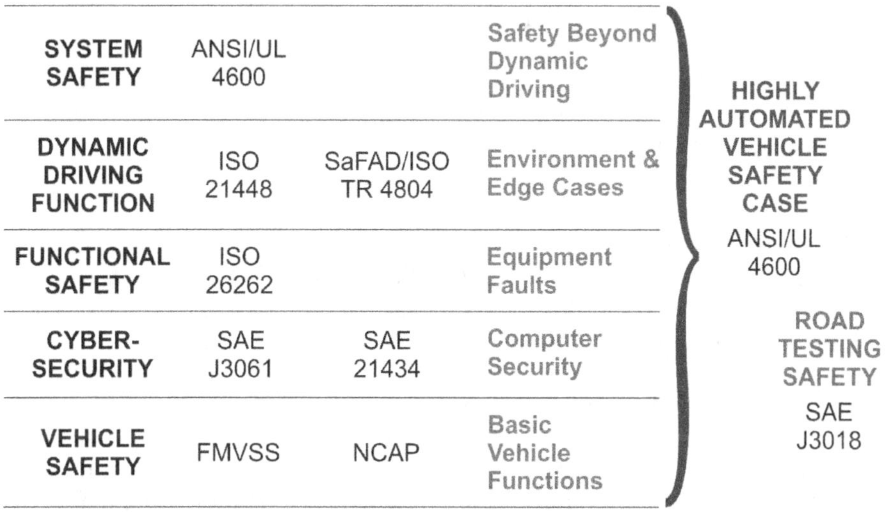

SYSTEM SAFETY	ANSI/UL 4600		Safety Beyond Dynamic Driving	HIGHLY AUTOMATED VEHICLE SAFETY CASE ANSI/UL 4600
DYNAMIC DRIVING FUNCTION	ISO 21448	SaFAD/ISO TR 4804	Environment & Edge Cases	
FUNCTIONAL SAFETY	ISO 26262		Equipment Faults	
CYBER-SECURITY	SAE J3061	SAE 21434	Computer Security	ROAD TESTING SAFETY SAE J3018
VEHICLE SAFETY	FMVSS	NCAP	Basic Vehicle Functions	

Figure 2.1. An illustrative organization of how automotive and AV standards can work together under UL 4600.

2.3. Frequently asked questions and misconceptions

A number of questions, misconceptions, and other topics commonly arise in discussions of UL 4600.

Q: Where can I get a free copy of the standard?
A: See Chapter 18 of this book for more information related to the standard, including how to access a free copy for online viewing directly from ULSE.

Q: Isn't it too early for an AV safety standard?
A: No. While locking the industry into quickly outdated technology is a legitimate concern, UL 4600 addresses this concern in two ways. First, it standardizes evaluation criteria for a safety case rather than the way an AV itself must be built. Those criteria are open-ended and have significant flexibility. Second, the ULSE standards process can operate on a very quick timeline (much quicker than ISO standards for example), adapting future

versions of UL 4600 to fix any unforeseen problems for AV makers much more quickly than an automotive design cycle takes. In principle, an urgent change might be made in a few months, not the 5 years (or more) it might take with other standards processes that have mandated multi-year change cycles. At the same time, the UL 4600 STP is mindful of the pitfalls of a high standards churn rate. Provisions are included in UL 4600 to permit a graceful transition after a new edition comes out, and the STP considers the question of impact on current standard users when considering changes.

Q: Does UL 4600 replace or conflict with ISO 26262 and/or ISO 21448?
A: No. UL 4600 was explicitly designed to work with ISO 26262 and ISO 21448. That goal was supported by multiple STP members and other stakeholders who also serve on technical committees for ISO 26262 and ISO 21448.

Q: Does UL 4600 apply to SAE Level 3 vehicles?
A: Yes, the automated driving function of Level 3 vehicles is clearly within scope for UL 4600. UL 4600 additionally covers vehicle safety responsibilities for the handoff process to the human driver.

Q: Does UL 4600 require a 3rd party assessor?
A: No, there is no requirement whatsoever for a 3rd party assessor. The independent assessor must be "independent" in a practical sense to ensure an unbiased review of the safety case. But there is no requirement for that independent assessor to work for a different company than the design team.

Q: Does UL 4600 require disclosing proprietary information publicly?
A: No, there is no such requirement. A conformance statement should show that a thorough assessment has taken place, what item the assessment applied to, and any unresolved discrepancies. However, there is no obligation to disclose proprietary design information nor any details from the safety case. An external assessor, if one is used (which is *not* required), needs access to details, but can be bound by a non-disclosure agreement without disrupting the conformance assessment process, as is commonly done when establishing conformance to other safety standards.

Q: Is UL 4600 an academic proposal or one company's guidance vs. a "real" standard?
A: UL 4600 is a real American National Standards Institute (ANSI) standard. As with many standards, UL 4600 started as a proposal. In this case the proposal author has significant industry experience and also works at a university. However, that initial submission went through multiple rounds of discussions, comments, editing, revisions, and a consensus process involving many dozens of stakeholders. Those stakeholders included vehicle manufacturers, automotive suppliers, chip makers, standards experts, consumer advocates, state regulators, federal regulators, international

Overview and applicability of UL 4600

representatives, researchers, and other safety experts. It met the rigorous process and consensus requirements to be issued by UL Standards and Engagement (ULSE) as both a UL and ANSI standard. ULSE is an accredited standards development organization that entirely controls the content, subject to a consensus-based governance process by the UL 4600 Standards Technical Panel. This is a bona fide, real industry standard.

Q: Is UL 4600 just a US standard?
A: Not really. While ULSE issues ANSI standards, there was significant involvement in the consensus process from not just North America, but also Europe and Asia. In time, it seems likely that other countries will adopt UL 4600 as a national standard, especially in Asia. UL 4600 has benefited from extensive international feedback, and definitely encompasses an international perspective in creating AV safety cases.

Q: Do I have to pay UL to be assessed for conformance?
A: No. ULSE does not sell conformance services. A different company, UL Solutions, along with a number of other accredited assessment organizations, might offer conformance services. However, there is no obligation to use any 3rd party assessment organization at all to establish conformance to UL 4600. Conformance assessment can be done completely in-house so long as arrangements are made to keep in-house assessors suitably independent.

Q: How do I actually make a safety case?
A: UL 4600 does not specify any procedure for creating a safety case. There are different possible approaches, and the safety case for each AV is likely to differ with the operational design domain, operational profile, equipment choices, and functional organization of the company building and deploying the equipment. Guidance for creating safety cases will need to come from other sources. UL 4600 does not tell you how to create a safety case, but rather provides a way to be confident that any safety case you have created is fit for purpose.

Q: UL 4600 is weak because it does not provide an objective, defined test that would prove safety. Why should we use it?
A: This common criticism is based on a false premise. It is not possible to demonstrate the safety of a non-trivial software system via testing alone. Testing is part of ensuring safety, but by far not the only part. Like other functional and software safety standards, UL 4600 helps the design team tell a story of how engineering rigor combines with testing to ensure safety. Moreover, UL 4600 permits the design team to only spend resources on testing that will contribute to ensuring safety, rather than mandating an arbitrary test regime that might actually prove nothing useful about safety. This flexibility is a strength rather than a weakness.

Q: But UL 4600 is so very long! How can I possibly conform?

A: ISO 26262 is several times longer, and applies to any public road vehicle, not just AVs. If you can conform to that, UL 4600 is a comparatively small incremental conformance effort. Everything in UL 4600 is there because it matters for AV safety. To be safe you'll need to do all those things, either with or without the standard. At least with the standard you don't have to reinvent the wheel and/or risk missing a safety-critical issue.

Q: How can I contribute to UL 4600?

A: Any interested party can request to be added as a stakeholder to the UL 4600 development process. This confers an ability to propose changes and comment on proposed changes. Interested parties can also participate in groups that work on proposed updates. However, stakeholders do not get voting rights. Participating as a voting STP member requires a formal application process, but involves no membership fees and is open to different types of stakeholders consistent with maintaining a reasonable balance of different stakeholder types. Application can be made by contacting ULSE. As of the time of publication of this book, you can request to be either a stakeholder (the usual request) or an STP member (for highly qualified candidates; formal application required) by contacting Ms. Deborah Prince of ULSE at deborah.prince@ul.org

2.4. Resources

- ISO 21448: https://www.iso.org/standard/77490.html
- ISO 26262: https://en.wikipedia.org/wiki/ISO_26262
- ISO/PAS 8800: https://www.iso.org/standard/83303.html
- MIL-STD-882: https://www.sebokwiki.org/wiki/System_Safety
- SAE J3016: https://users.ece.cmu.edu/~koopman/j3016/
- SAE J3018: https://www.sae.org/standards/content/j3018_202012/
- TR 4804 (pre-standard SaFAD report version):
  https://www.aptiv.com/docs/default-source/white-papers/safety-first-for-automated-driving-aptiv-white-paper.pdf

- UL 4600 does not provide an answer to the question of how safe might be "safe enough." That question is discussed in Koopman, "How Safe Is Safe Enough," 2022: https://safeautonomy.blogspot.com/2022/09/book-how-safe-is-safe-enough-measuring.html

3. Requirements and prompt elements

Summary: Section 4.1 of UL 4600 describes the structure of the standard, and in particular the use of prompt elements. Each prompt element identifies a concept, activity, hazard, or other topic that should be addressed by the safety case. A prompt element classification scheme permits conditional deviation for most prompt elements to provide flexibility.

Those who have read many standards will recognize that UL 4600 has a distinctly different style than a typical standard. The biggest difference is in the use of prompt elements.

3.1. Requirement structure

The portions of UL 4600 that have to do with the contents of the safety case and conformance process (clauses 5-17) are structured as a set of requirements supported by prompt elements. Figure 3.1 is a representative excerpt from a draft of the first edition. Every requirement in UL 4600 follows the general format shown in fig. 3.1:

- A stated requirement in the form of a "shall" statement
- An enumerated list of Mandatory prompt elements
- An enumerated list of Required prompt elements
- An enumerated list of Highly Recommended prompt elements – in this case empty, as indicated by "N/A" (Not Applicable)
- An enumerated list of Recommended prompt elements – in this case empty, as indicated by "N/A" (Not Applicable)
- A conformance process statement
- Optional examples located anywhere in the requirement structure
- Optional references and notes located anywhere in the requirement structure
- Optional numbered notes that apply to the whole requirement structure (not shown, but would start with numbering 12.3.1.6.1. for fig. 3.1)

Each requirement (12.3.1 in fig. 3.1) is a classical "shall" statement regarding some property of the safety case, an activity that must be documented by the safety case, or the like. In this case, the requirement has to do with the safety case documenting verification & validation coverage of faults.

The requirement statements are quite general, and provoke the question of "what exactly might that mean in detail?" Here is where UL 4600 diverges from the style of a more traditional standard. Rather than having dozens of supporting "shall" requirements, it instead provides a list of prompt elements.

12.3.1 V&V shall provide acceptable coverage of safety related faults associated with the design phase.

12.3.1.1 MANDATORY:
   a) Systematic design defects
   b) Design consideration of faults, corruption, data loss, and integrity loss in sensor data
   c) Requirement gaps/omissions and requirement defects
   d) Response to violation of requirement assumptions
      EXAMPLE: Response to exceptional operational environment
   e) Identification and description of the intended ODD
   f) Acceptable mitigation of aspects of the defined fault model for each component and other aspect of the item

12.3.1.2 REQUIRED:
   a) Maintenance procedure definitions
      NOTE: While maintenance occurs during the lifecycle, the definition of procedures needs to correspond to design requirements and assumptions made in design regarding maintenance.
   b) Operational procedure definitions (including startup and shutdown) and operational modes
   c) Faults, corruption, data loss, and integrity loss in data from external sources
   d) Faults and failures associated with exceptional conditions that impair risk reduction functionality
   e) Hardware and software errata and other third-party component design defects
   f) Other faults in safety related functions, component designs, and other designed properties

12.3.1.3 HIGHLY RECOMMENDED – N/A

12.3.1.4 RECOMMENDED – N/A

12.3.1.5 CONFORMANCE:
Conformance is checked via inspection of design and V&V evidence.

Figure 3.1. Excerpt from UL 4600 first edition voting draft. Copyright © 2019 ULSE Inc., used by permission.

3.2. Prompt elements

A prompt element is a phrase one might expect to see in a bullet list of details supporting the requirement. The term "prompt" is used because it is in the style of a memory-jogging prompt to remember to include something in the safety case. Loosely, every prompt element is "did you think of that?" within the context of its parent requirement.

The structure of using prompt elements saves words by avoiding long, tediously structured requirements statements. For the example in fig. 3.1 instead of saying "V&V shall provide acceptable coverage of safety related systematic design faults associated with the design phase," the prompt element 12.3.1.1.a. simply says "systematic design faults." The "V&V shall provide acceptable coverage…" part of that requirement is implied by its position supporting requirement 12.3.1.

The approach of a prompt element is that a reasonable engineer would naturally appreciate how that prompt fits in within the overall requirement. The prompt element might provide specific scope included in the requirement, elaborate aspects of the requirement that might (or might not, depending on the engineer's experience) have been obvious to include, and so on.

In some cases there might be reasonable ambiguity or uncertainty as to what a prompt element might mean. Sometimes there is a note to provide guidance or additional considerations. But more often, prompt elements are clarified by providing non-limiting examples.

Safety cases should include all relevant issues that might affect safety, with prompt elements being a list for getting started rather than a boundary that prevents relevant issues from being included in the safety case. The prompt elements, and especially any included examples, are discussion-starters for the safety case, not limits on what can be included.

From a standards interpretation point of view, requirements and prompt elements are a normative minimum of what needs to be addressed by the safety case (they have to be considered). Notes are intended to be rationale or other supportive text that guides the interpretation of the requirements and prompt elements.

Examples serve two main purposes. The first is that they help illustrate what might be meant by the prompt element in a definition-by-example approach. The second is that they provide grist for the safety case mill in terms of capturing potential lessons learned and secondary "did you think of that?" considerations. While the standard does not require any particular example to be addressed in the safety case, an assessor should use the example list as a starting point to consider whether the prompt element has been addressed thoroughly enough for acceptable safety.

In practice, if an example is an obvious fit for the safety case, it should be addressed. If an example does not apply, nothing further need be said in the safety case. Additionally, other considerations relevant to the prompt element that matter to the particular safety case should be addressed. Some additional considerations might be obvious extrapolations of the examples. Other considerations might be problems the design team has encountered before, or any other issues they think of that matter for ensuring acceptable safety.

The boundary between two overlapping prompt elements might be a matter of taste. In those cases, a portion of the safety case might trace to either (or both) prompt elements without concern. The point of prompt elements is not to stir debates as to which bin a particular hazard might fall into, but rather to make sure that no hazards slip through the cracks.

It is important to keep in mind that not every prompt element needs to be associated with a technical mitigation. Some hazards might be mitigated by a specific software function. But other hazards might be mitigated using well-documented and audited maintenance procedures. Yet other hazards might simply be declared acceptable for the particular design and operational environment of the AV under consideration because they will not disrupt AV operation, or are considered so rare that they present a negligible threat to safety. Addressing a prompt element does not necessarily mean creating a component or software function. Rather, addressing a prompt element requires including it in the safety case and explaining how it has been addressed (or why it need not be addressed, assuming such a deviation is permitted).

At the end of every requirement is a statement of the applicable conformance procedure. Usually this explains what work products are checked by assessors. In future versions of the standard this might become more specific as lessons are learned about possible issues that arise in AV safety cases.

3.3. Deviations

Some standards permit the concept of tailoring, in which design teams can elect to disregard portions of the standard when considering conformance, or relax rigor that would theoretically be required. While this might still result in a safe enough product if tailoring is done in a thoughtful way, it might also result in corners being cut that compromise safety outcomes. This traditional type of tailoring is supplanted in UL 4600 with a more formalized – but still flexible – deviation mechanism.

Every AV safety case is likely to have different needs, and it is unreasonable to think that every single prompt element will apply to every single AV. This is especially true when considering a potential range from sidewalk delivery robots (that nonetheless still must cross roads) to passenger vehicles to heavy transcontinental freight trucks. There needs to be a controlled way to say "this does not apply" in a safety case, but without opening the floodgates to simply delete every prompt element that seems inconvenient, especially if the potentially deleted prompt elements are essential to safety for all vehicles.

UL 4600 provides a structured deviation framework that allows a safety case to designate prompt elements as not applicable without disregarding crucial parts of the standard wholesale. The complete description of the deviation process is described in UL 4600 section 4.1.1, and is summarized as:

- All **requirement statements** shall be fully addressed in the safety case (e.g., 12.3.1 in fig. 3.1.).
- All **mandatory** prompt elements shall be fully addressed in the safety case (e.g., 12.3.1.1 in fig. 3.1.).
- All **required** prompt elements (e.g., 12.3.1.2 in fig. 3.1.) shall be fully addressed, unless they are "intrinsically incompatible" with the item. This has the net effect of being mandatory except when the very nature of the item makes it inapplicable due to its construction, operation, lifecycle, or other inherent system characteristic. As a simplistic example, if an AV design does not have any hydraulic systems, then prompt elements regarding hydraulic faults can be traced to a deviation

Requirements and prompt elements 17

statement rather than an elaborated argument, with the statement being something like: "deviation: does not apply / no hydraulic components." However, if there is any way to show relevance of the prompt element to the item, then it must be addressed.

- **Highly recommended** prompt elements (e.g., 12.3.1.3 in fig. 3.1) should be met, but deviations are permitted with a plausible rationale included in the safety case. That rationale must be tracked for continuing acceptability as part of impact analysis for any safety case change. As an example, consider a highly recommended prompt element referring to software unit test. A team might say that their peer review process is so effective that unit testing is a waste of resources and choose not to do unit testing at all. This rationale is plausible, so such a deviation might be acceptable. However, this deviation justification amounts to a sub-argument in the safety case that hinges on measuring peer review effectiveness. The argument that peer reviews are so effective that unit testing is a waste of time might be falsified by data showing peer reviews have become ineffective at detecting defects that would reasonably have been caught by unit testing.
- **Recommended** prompt elements (e.g., 12.3.1.4 in fig. 3.1) are suggestions that might be useful to some teams, but need not be mentioned in the safety case at all. Deviations are both unlimited and impose no documentation burden in the safety case. One might imagine that recommended prompt elements could, over time, be promoted to highly recommended prompt elements in future versions of the standard if the community's experience shows they have consistent value in assuring safety.

Summarizing, requirement statements and mandatory prompt elements must be fully addressed by the safety case. Required and highly recommended prompt elements must also be addressed, but deviations are permitted with a technically substantive explanation recorded in the safety case. Required prompt element deviations are only permitted when the prompt element simply does not make sense given the item that is the subject of the safety case. Highly recommended prompt element deviations are more permissive, but still require a convincing story. Recommended prompt elements are entirely optional.

There is a special case for Elements Out Of Context (EOOC), described in this book in section 5.2.3. An EOOC amounts to a safety case fragment that is maintained separately, and in principle not disclosed to the maintainer of the main AV safety case. An EOOC might cover a hardware component, software component, cloud computing service, or some other aspect of the safety case that is not under the control of the AV design team. An EOOC might not satisfy all prompt elements, including not addressing some mandatory prompt elements in UL 4600. However, the EOOC interface must export an obligation to resolve any omitted prompt elements to the user of the EOOC. This means that the EOOC interface tells the AV design team what prompt elements are not covered by the EOOC's hidden safety case, and imposes a burden on the AV design team to cover them in their safety case instead.

4. Terminology

Summary: Section 4.2 of UL 4600 provides a number of definitions. In this section we summarize some of the key conceptual terms.

Getting terminology right is both tricky and time consuming for any standard development effort. Partly that is because precision is required, and partly that is because terminology choices must take into account existing definitions from other relevant standards. Those existing definitions might not quite be a fit, might be limited in ways that cause problems unforeseen by the original definers of the term, or might even plainly contradict each other across standards.

The definitions section changed significantly from the first to second editions of UL 4600. The intended usage of terms did not change in any major way. Rather, the definitions were revisited to improve clarity and reconcile any differences with other standards using the same or similar terms. The definitions section is expected to stay generally stable for subsequent versions, such as the pending 3rd edition.

4.1. Terminology approach

In the 2nd edition update, the decision was made to avoid any conflicts with the essence of any ISO 26262 and SAE J3016 defined terms, even if other standards might use the term in a different way. What this means is that if UL 4600 defines a term, it means the same general concept as ISO 26262, but might have additional scope.

The additional scope approach is implemented as a "strict superset" policy. As an example, the ISO 26262 definition of "assessment" is limited to assessing objectives within ISO 26262. UL 4600 needs the term "assessment" to additionally apply to the objectives of UL 4600. Picking a different term for "assessment" would simply confuse things, especially since a casual reader might not even realize this limitation is built into the ISO 26262 definition. So UL 4600 expands the scope of the term "assessment" to also apply to the safety case, using the term in a compatible way. This expansion of scope explicitly applies to the terms "assessment" (defined in ISO 26262), "operational design domain" (defined in SAE J3016), and "safety case" (defined in ISO 26262).

For other terms, a minimalist approach was taken to providing definitions. A number of terms already defined by ISO 26262, MIL-STD-882, or other standards were used without providing specific definitions, although sometimes with an implicit larger scope that should be obvious to a reader. For example, some terms defined in ISO 26262 explicitly include a limitation to "functional safety" or specifically refer to ISO 26262 work products. The same term used in UL 4600 is intended to encompass system safety and work products that might go beyond the scope of the strictest interpretation of that term in ISO 26262, to also encompass UL 4600-related work products and system safety.

Other terms are used in the way one might expect an experienced automotive or autonomous vehicle engineer might use them. The informal test applied was that if a web search turned up uses of a term compatible with the intended use on the first search result page, there was no point in including the definition in UL 4600. In cases where the design team feels there is ambiguity, any reasonable specific definition for a term can be used so long as that team's definition is included in the safety case.

4.2. Key conceptual terms

The terminology definitions in UL 4600 follow the so-called "ISO substitution rule." That means that, at least in principle, any defined term can be replaced with the definition text without changing the meaning. This explains why UL 4600 terminology definitions do not start with a capital letter and do not end with a period. The defined terms are essentially macros that are conceptually expanded when reading the standard.

The official definitions are those given in the currently issued version of UL 4600. Rather than replicate the exact terminology language here, the following list highlights a few key terms that are central to understanding the standard.

One term used in this book that is not used in the standard is the abbreviation "AV" for autonomous vehicle. The standard uses the more correct term "item" instead (defined below). The difference is that an "item" is everything that contributes to the safety of an AV, regardless of whether it is on the vehicle or not. Just using the term AV raises ambiguity, especially in situations involving remote computational support for AV operation. "AV" is used in this book because the word "item" is foreign to anyone who is not steeped in the ways of ISO 26262, and would be too likely to confuse a reader, especially one who is just browsing a particular book section. Nonetheless, "item" is a better word to use in safety cases and more rigorous technical writing for audiences who have already been oriented to the terminology. When this book uses the term "AV" in a discussion, it really means "item" unless it is talking about the physical properties that are necessarily associated with an actual vehicle (e.g., vehicle speed).

This list is non-exhaustive, and intended to provide conceptual guidance rather than precise definitions.

- **Acceptable.** The goal of UL 4600 is to determine whether a safety case is acceptable. Additionally, the terms "acceptable safety" or "acceptably safe" are often used. The key notion here is that some objective definition of how safe is "safe enough" has been stated by the design team, likely as context for the top-level safety case claim. That will end

up being something like "The AV will be acceptably safe" as a top-level claim, with "acceptably safe" defined in a supporting context note. So the notion of "acceptable" is not a subjective opinion of the design team, nor an assessor. Rather something being "acceptable" means that it contributes to the safety case meeting specific safety criteria defined in the safety case.
- **Argue.** This is a shorthand term for building a safety case that consists of a set of claims, arguments, and supporting evidence. One might "argue" that a claim is true by building a subset of a safety case to support the claim. As with "acceptable" this is not a subjective concept, nor a rhetorical exercise, but rather a reference to a subset of a sound safety case.
- **Autonomous.** This describes the behavioral capabilities of an item covered by UL 4600. This differs from the term "automated" used by J3016 primarily in that J3016 talks about "automated driving systems" (i.e., just the functionality of driving the vehicle). In contrast, the scope of UL 4600 is all functionality of the vehicle and its lifecycle support (the "item"), which goes far beyond just the driving function. An autonomous vehicle does not require a human driver to continuously monitor it to achieve acceptable safety.
- **Conformance.** A safety case conforms to UL 4600 if it meets all the requirements in the standard, including having undergone both self-assessment and independent assessment.
- **Element Out Of Context (EOOC).** This is a fragment of a safety case that is not visible to the main item safety case. A contract-based design style EOOC interface is provided as a stub in place of a component's detailed safety case. A component vendor might, for example, provide an EOOC interface for a sensor component or a real time operating system that can provide sub-claims to be built on by the item's safety case. This is analogous to an ISO 26262 Safety Element out of Context (SEooC).
- **Incident.** This is any situation in which the vehicle exhibits unsafe behavior or does not meet the claims of the safety case. A near-hit situation is an incident, as is a vehicle crash. An improperly mitigated internal fault is also an incident, even if no adverse vehicle-level behavior results from that situation. Any incident needs to be analyzed to determine the root cause and identify any appropriate corrective action.
- **Item.** This is the scope of the system-level safety case. Typically that will include not only the vehicle, but also any support infrastructure, development processes, and lifecycle support activities. The item encompasses everything relevant to safety. The safety case must address the entirety of the item. (This terminology is adapted from the use of the term "item" in ISO 26262.)
- **Mitigate.** This means applying some approach (technical, procedural, operational, product definition, etc.) to reduce an unacceptable level of risk to an acceptable level of risk. A risk might be "accepted" if it is

judged no mitigation effort is required to achieve acceptable safety, rendering that risk mitigated as well.
- **Operational Design Domain (ODD).** This is an abstraction of the environment that the item is intended to operate within. It is not the actual environment, but rather in some sense an engineering "model" of the environment that matches the intended subset of the real world the AV is intended to operate within. (The term "model" is used here in a very general sense in that a simulation is also a model of world and road users, and not necessarily in the stricter sense of a model created in a model-based design workflow.)
- **Risk.** This is a combination of probability and severity, which is a commonly used definition. Users of ISO 26262 will want to address the ASIL-related concept of "controllability" when considering risk, taking into account how controllability might be apportioned to the AV's autonomous driving functionality.
- **Safety Performance Indicator (SPI).** This is a metric that is related to safety rather than other aspects of performance. Think of it as a Key Performance Indicator (KPI), but one that must have a very direct relationship to safety, and therefore should be traceable to a specific claim in the safety case. Chapter 16 discusses this in more depth.

Some other terms commonly used by UL 4600 will be defined in the course of later discussion:
- Chapter 5: safety argument, safety case, claim, argument, evidence.
- Chapter 6: hazard, risk
- Chapter 16: field engineering feedback, more on SPIs
- Chapter 17: assessment, assessor, self-assessment, independent assessment

5. The safety case

Summary: Clause 5 of UL 4600 requires a safety case that includes claims supported by argument and evidence. The safety case must encompass all aspects of safety, including difficult-to-reproduce aspects of the item. Areas that are called out for special attention include accepted risks (risks that are not fully mitigated) and safety culture.

5.1. Structure of a safety case

UL 4600 uses a form of the usual definition for a safety case: a "structured argument, supported by a body of evidence, that provides a compelling, comprehensible and valid case that a system is safe for a given application in a given environment." (This definition is taken from *Defence Standard 00-56 Issue 7 (Part 1): Safety Management Requirements for Defence Systems*, UK Ministry of Defence, p. 26.)

Breaking this down, at the minimum a safety case will have the following elements:

- **Claims:** Statements regarding safety or related properties that are held to be true within a stated context. For example, a mid-level claim might be "AV does not collide with pedestrians." If the claim is true, that portion of the safety case likely supports a top-level claim such as "AV is acceptably safe."
- **Argument:** Elements of the safety case that support the claims by presenting reasoning why the claim is thought to be true. In UL 4600 parlance, the safety case "argues" the claims are true via the inclusion of a supporting argument and evidence. An argument might be something like "sensors detect pedestrians soon enough so that the AV can maneuver to avoid collisions with them." That argument is, in turn, supported by sub-claims that might cover whether sensors can actually detect pedestrians and whether the AV is actually able to maneuver to avoid pedestrian collisions.
- **Evidence:** Information that supports an argument. For example, a set of simulated scenarios might show that sensors did indeed detect pedestrians soon enough for the AV planning and control logic to avoid collisions. Other evidence might be the result of an engineering analysis, or process quality data.

Generally a safety case starts with one high-level claim and breaks it down into multiple sub-claims that are, after enough layers, small enough to have direct supporting evidence. Other elements should be included in the safety case, such as descriptions of strategies for decomposing broad overarching claims into more detailed, smaller claims.

A hypothetical safety case fragment is shown in figure 5.1. The figure is mostly in the style of a Goal Structuring Notation diagram, although with GSN a "claim" would instead be called a "goal." We use the term "claim" to be consistent with UL 4600's use of that term. UL 4600 does not preclude using "goals" in notation instead of "claims" if that terminology is preferred by a particular design team.

In fig. 5.1 the root claim at the top is "AV does not pass too close to pedestrians." This would itself be a sub-claim of some higher-level claims – ultimately leading up to a claim of overall AV safety at the top. But for this discussion we just consider this safety case fragment.

Context boxes are typically required to provide supporting information for interpreting the claim. In this example one context box explains what "too close" might mean. Another context box explains that the term "pedestrian" includes a broader class of people beyond just unassisted walkers. In a real safety case, these context boxes might need quite a lot of information to make claims unambiguous.

Keeping a claim simple enough to understand tends to involve shoving complexity into supporting context. This context might not need to be physically written in boxes for every claim. For example, there might be a separate list of defined terms such as "pedestrian," which can just be used in a claim without explicitly spelling out the details. But the details need to be somewhere in the safety case.

Below the root claim in figure 5.1 is a strategy box. This describes how the claim above the strategy box is being broken down into sub-claims. Overall the safety case is a tree-like structure (more generally a directed graph) with the root at the top and branches below that breaks big claims into smaller, easier-to-deal-with sub-claims. Alternating layers of strategies and sub-claims is one approach to what is collectively known as an "argument."

While safety cases are sometimes published with just the claims, that is likely to result in a safety case that cannot be properly evaluated. The reason is that for a complicated breakdown strategy it can be difficult for a reviewer to reconstruct the logic being used in the breakdown. That in turn might make it difficult to ensure no important sub-claim has been left out. A nested set of claims with arbitrary decompositions not explained as a coherent argument is unlikely to be a sufficiently well-stated safety case.

At the bottom of the safety case structure is evidence supporting all the claims. In practice, there is likely to be a specialized type of evidence sub-claim that summarizes the conclusion that should be drawn from some evidence. (Use of claim specifically designated as an "evidence claim" is not required by UL 4600, but seems like a useful practice.) A single set of evidence data might have multiple evidence claims summarizing different aspects of that data to support multiple different arguments within the safety case. The evidence might range from road testing data to simulation results to software quality data. Evidence might also be an expert opinion stated as such. The important part is that the evidence itself can be reviewed when judging the credibility of the corresponding evidence claim.

The safety case

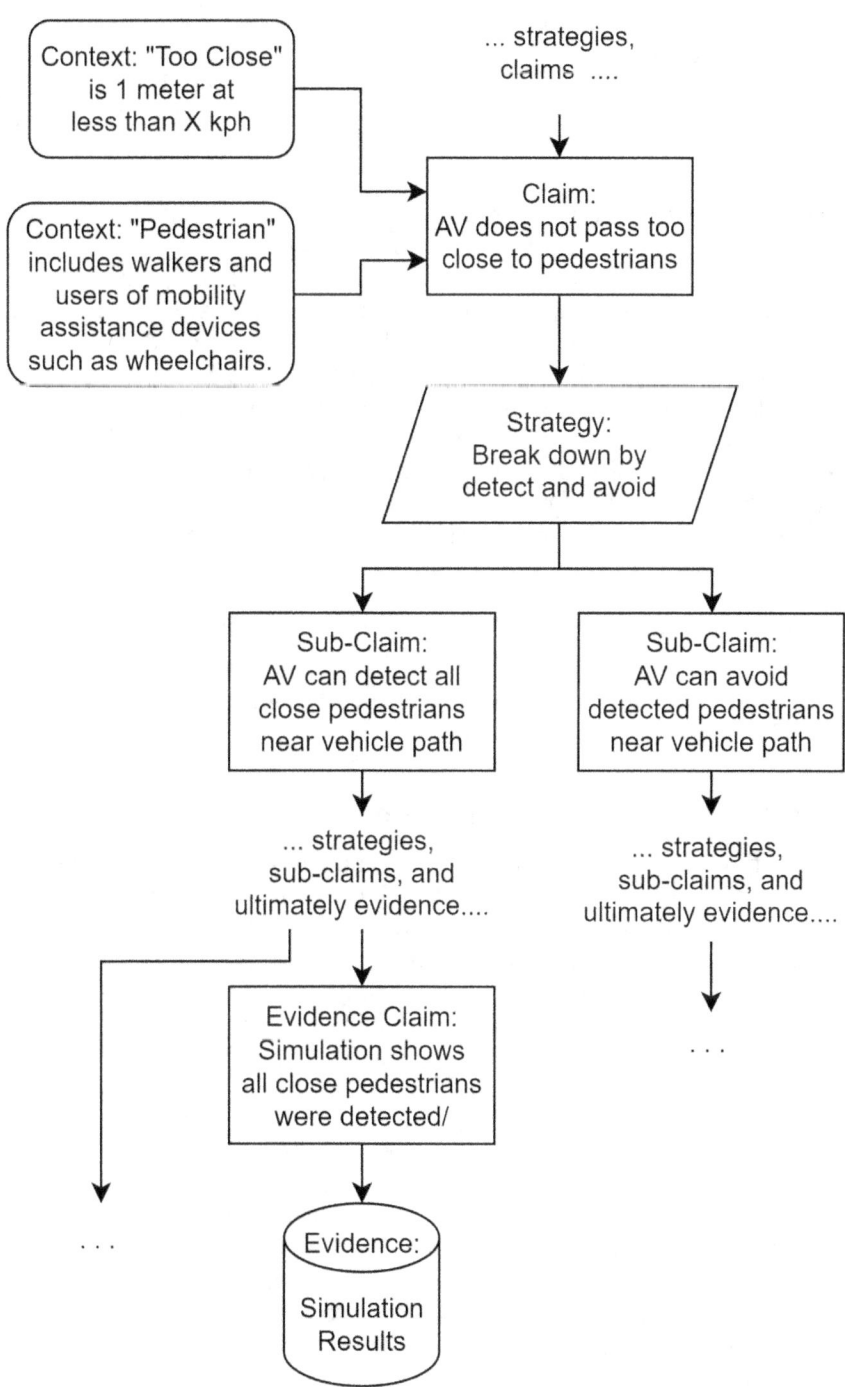

Figure 5.1 Example Safety Case diagram.

While safety cases are often drawn using a graphical notation, UL 4600 does not require doing that. Safety cases can be any combination of graphical notation, text notation, and whatever else makes sense so long as the meaning of the notation is well defined. Some possible established safety argument notations mentioned in the standard are the OMG Structured Assurance Case Metamodel (SACM), Goal Structuring Notation (GSN), and Claims Argument Evidence (CAE).

It would be no surprise if safety cases use a graphical notation at the top level and for selected portions of the safety case, but use other representations for other parts of the safety case. A different approach might be to use a purely text-based safety case description language for the whole safety case, then render portions of the safety case on demand graphically using tool support (e.g., "show me a graph of all the claims connected to this particular sub-claim" while not showing the entire safety case).

While in principle a freeform purely prose document might suffice as a safety case, the need for cross-referencing, traceability, and other factors make that seem a difficult approach in practice. Some sort of data structure or database organization of a safety case seems more likely to be viable.

5.2. Safety case properties

To be acceptable, the safety case must have a number of properties and have a sufficiently comprehensive scope.

5.2.1. Safety case scope and the item

The safety case must encompass all relevant safety considerations for the item being described. That includes not only components inside the AV, but also any critical roadway infrastructure, lifecycle support processes, online data services (including Internet cloud data centers), and any other aspects of operation and logistics that might impair safety. Taken together, this entire scope is called the "item" for which the safety case has been created. While it is possible that the item might just be an AV itself, in practice it is likely to be more expansive, including at least maintenance and lifecycle processes that are undertaken to ensure the AV's equipment is maintained properly. (The term "item" was chosen to be consistent with ISO 26262.)

A particular consideration for so-called "autonomy kits" that are added on top of conventional vehicles is that those vehicles might themselves have defects that the AV kit design team might be unaware of. Human drivers are expected to be able to compensate for at least some instances of vehicle malfunctions, so the autonomy kit would need to deal with that as well. The safety case would need to argue that malfunctions of the underlying vehicle could be compensated for by the autonomy kit.

Some aspects of the operational environment might be relevant for safety, but not under the control of the design team. These might include the quality of painted road stripes and the availability of connectivity to online data sources. To the degree that the safety case cannot argue such aspects are sufficiently dependable, it needs to instead argue that any failures can be tolerated safely.

An important characteristic of safety case scope is an invocation of the principle "if it isn't written down, it didn't happen." If something is not documented in the safety case, the safety argument is not allowed to take credit for it. This means the safety case must completely support the safety claims being made with argument and evidence. Any loose ends mean the safety case is incomplete, and therefore not in conformance with UL 4600.

UL 4600 section 5.3 puts a special emphasis on the safety requirements for the item being included in the safety case. That includes not only intended functionality, but also unintended functionality and potentially unsafe behaviors that might be exhibited by the AV in operation. The safety requirements must also address how faults are mitigated, including design faults, operational faults, abnormal operating environment conditions, and lifecycle issues.

In short, the safety case needs to include everything that might pose a hazard or impose risk, and also include how those potential problems have been mitigated to result in net acceptable safety.

5.2.2. Concept of operations

A required element of the safety case is a description of the concept of operations for the item. That includes a description of capabilities, safety objectives, risk mitigation strategy, architecture, operating modes, and operational design domain (ODD).

The purpose of this requirement is to give assessors an overview of the system that is being addressed by the safety case. Any good review requires the assessor to have some understanding of what is being assessed, and this requirement addresses that need.

5.2.3. Element Out Of Context interfaces

There is also a requirement to clearly identify any boundaries with other systems that are not included within the bounds of the safety case, but are relevant to safety. UL 4600 includes the concept of an Element Out Of Context (EOOC). An EOOC is a system element that has a defined safety interface for use by the item safety case, but for which details of the safety case are not available for assessment.

The idea is that some component vendor (perhaps a real time operating system vendor, or a sensor company) provides a system element to the AV system integrator. The system integrator needs to make safety claims based on the capabilities of that component, but does not have access to the component's safety case. The EOOC interface provides a clean boundary with claims about the component that can be used by the AV system integrator. The arguments for those claims are not provided, but will have been assessed according to UL 4600 in a separate process by the component vendor.

The EOOC interface also provides an unambiguous statement about which aspects of UL 4600 are not satisfied by the component's hidden safety case. The system designer must find a way to satisfy those unresolved prompt elements within the system safety case since they are not handled by the components safety case. For example, a sensor maker might provide technical data about how to perform calibration, but leaves it to the system integrator to address how to ensure calibration is performed correctly and when needed in the system-level safety case. (This interface approach will be familiar to those who have worked with a microcontroller vendor who has provided a safety manual for their chip.)

5.2.4. Types of claims

There are potentially several different flavors of claims that might be included in a safety case. UL 4600 does not really distinguish which types of claims need to be made where, and does not give names to different types of claims. Nonetheless, it can be helpful to informally describe the variety of claims that might be used in a safety case here.

- **Performance claims** are perhaps the most obvious, and would be made about how well different safety-related functions accomplish their tasks. For example, a claim might be made about what fraction of relevant pedestrians is detected by a sensor fusion system.
- **Mitigation claims** specifically deal with a hazard or risk that has been identified, and summarize an argument that the risk has been mitigated. All risks need to be covered by safety case claims one way or another, although some risks might be accepted rather than fully mitigated (discussed later in this chapter).
- **Quality and dependability claims** might be made about failure rates of equipment or defect rates in design. For example, a claim might be made about the mean time between safety-relevant failures for a sensor device, or how often a software defect is likely to result in a dangerous behavior.
- **Design property claims** might be made about various characteristics of the design. For example, a claim might be that there is no substantive common cause failure mechanism for two redundant computing elements that need to fail independently to ensure safety.
- **Validation claims** might be made in support of performance, quality, or design claims. For example, the result of a set of tests might show performance outcomes in support of a performance claim, or help substantiate a claim of no safety-relevant software defects being found via a defined type and amount of testing.
- **Process quality claims** might be made in support of other claims. For example, a claim that software defect rates are low, or that a component has an acceptable level of safety integrity will need to be at least partly based on a claim that the engineering process has high quality.

The safety case

- **Operational environment claims** might be used to support arguments that a particular issue is not relevant. For example, a claim that the system will not encounter rain in a particular operational area with a sunny-day-forecast-only operational profile might be used to support an argument that sensor issues due to rain are not relevant.
- **Lifecycle claims** might be used to support an argument that some hazards have been mitigated via support procedures. For example, a periodic scheduled maintenance procedure might mitigate hazards related to loss of calibration if the maintenance procedure is defined properly and is checked to actually be performed.

There are likely many different flavors of claims beyond these. The point here is to illustrate that many claims in a safety case are likely to be of a different type than "system has behavior X" type functional claims, especially when creating supporting arguments and sub-claims.

5.3. Safety case tool support

Quite a lot of information will need to be in the safety case, perhaps with details in auxiliary documents referenced by a primary safety case. Additionally, there are many properties that safety cases must fulfill to be conformant. This inevitable complexity likely means there will need to be tooling support to implement UL 4600 safety cases in practice.

While the types of tools that are worth using will depend on circumstances and the complexity of any particular safety case, some likely candidates are:

- **Static analysis.** Safety cases are likely to involve a tree data structure that has some shared sub-trees (a tree structure for which some branches are allowed to reconverge, as might happen when several claims ultimately rely on the same pool of evidence or the same process quality claims). A static analysis tool can check to make sure there are no dangling pointers in the tree, that every claim is supported by an underlying argument and/or evidence, and so on. Just as static analysis tools such as compiler warnings can flag potential problem areas automatically, a safety case static analysis tool can point out when the argument is malformed without wasting human reviewer time to do such checks.
- **Conformance traceability.** UL 4600 has a large number of prompt elements that must be checked for conformance. AVs are complex, life-critical systems, so this should be no surprise. But tooling can ease the burden by making it easier to trace prompt elements to parts of the safety case. As a basic capability, an assessor might ask the tooling "show me all parts of the safety case that are relevant to conformance for 5.2.1" and the tool should provide links to the parts of the safety case that document the defined safety case formats as well as evidence that the formats have been followed as required by section 5.2.1 of UL 4600. There are a

number of crosscutting properties that need to be checked, and tooling support can help with this.

- **Browsing tools.** A full-size safety case is likely to be difficult to navigate without some support tooling. At the very least a search capability should be provided, preferably with a pointer mechanism so one viewer can send a "follow this link" descriptor to someone else for viewing just as one might do with an intra-net company Web page. Other capabilities such as dynamic selection of nodes based on a topical tag or what they reference might be desirable as well (e.g., "show me all claims that refer to this sub-claim," "show me all claims that relate to maintenance procedures," or "show me all arguments that are still missing evidence").
- **Impact analysis.** Any time a change is made in the safety case it is likely that related claims and argument structures will be affected and possibly also need to be changed. Tooling support could automatically mark a specific node in a safety case as having been changed, as well as automatically marking other connected parts of the safety case as a minimum set of areas that need to be reviewed for potential effects of that change. Similar capabilities could be used to support assessment. Tooling might both ensure that all safety case nodes are checked during initial assessment, and track that the effects of any changes have been reassessed after a change.
- **Natural language checkers.** Spell checkers, grammar checkers, and other similar checking tools for natural language can help avoid flaws in the natural language aspects of the safety case, just as they do in other written materials. Such checkers might also highlight risky argument phrasing such as "essentially all" or "should be" in a claim, call out potentially confusing use of double negatives, and so on.
- **Dashboard tools.** Managers might wish to have dashboard displays of progress in creating the safety case, size, complexity, number of static analysis issues that need to be fixed, number of nodes of the safety case that need to be re-assessed based on impact analysis after a change, and so on.
- **Safety performance indicator (SPI) tracking.** UL 4600 requires the use of SPIs to track data that might falsify claims. Tooling will be required to collect, process, and detect SPI threshold violations.
- **Viewers and import/export:** It is unlikely that all material relevant to a safety case is available and stored in one tool, or possibly even on one server system. The ability to link, view, and import/export data will be important to provide a reasonable experience to understand the safety case and see related design artifacts. Consistency across multiple copies of data is always difficult, so embedded live viewers might be used when practical to avoid version skew in imported/exported data.
- **Configuration and change management.** The safety case is a work product that needs to be managed just like training data, source code, test

results, and other work products. Changes to the safety case need to be tracked with a version control system. Each software release needs to have configuration management that associates it with a specific version of the associated safety case that was conformant with UL 4600 at the time the configuration was released for production/update. This is important not only for tracking the safety case version itself, but also for dealing with versioning considerations for safety performance indicators that might have had different metric and threshold definitions for different safety case versions.

5.4. Argument sufficiency

Just because something is in a proper format for being a safety case does not mean it is a credible explanation of why an AV is expected to be acceptably safe. The argument needs to be valid (well formed) and sound (both valid and logically true).

A valid argument is one that fully supports its conclusion. That primarily means it is not missing pieces, both obvious and subtle. If a claim has no evidence and no underlying argument, it is malformed and therefore invalid. A more subtle problem is an argument that is not obviously malformed, but has defects in its reasoning such as an incomplete set of sub-claims, or outright logical defects in an argument such as direct contradictions.

Even if an argument is valid (both well formed and complete), it might be unsound if the evidence shows that a claim is false. One hopes that this would not be the case at the time the safety case is assessed. However, it might be that a safety case is initially sound – but then later found to be unsound in light of some newly encountered evidence or some change in real-world circumstances.

Teaching the art of creating a great safety case is well beyond the scope of what can be accomplished in this book. (To be clear, that art is still evolving.) However, there are a number of attributes and pitfalls to be considered when creating a robust safety case that we can discuss. Some of them are more of the nature of a logic and argument college course, but ones specific to UL 4600 are discussed in the below subsections.

5.4.1. Inapplicable prompt elements

Not all prompt elements apply to all AVs for which a safety case is being developed. However, the safety case must record the disposition of all prompt elements to leave a paper trail. It is not acceptable to simply leave most prompt elements out without a mention. Rather they need to be addressed in the safety case, with any deviation meeting the UL 4600 deviation criteria with an appropriate argument for why a deviation is acceptable.

Summarizing Table 4.1 in UL 4600, mandatory prompt elements must be addressed. Required prompt elements that are not addressed must have an argument justifying that they are fundamentally inapplicable. Highly recommended prompt elements need a substantive rationale. The only prompt elements that can be omitted from the safety case with no mention are those that are categorized as recommended prompt elements. (See section 3.3 of this book for more on deviation rules.)

5.4.2. Assumptions

Ultimately there will need to be assumptions made to build a safety case, with each assumption being more formally called a premise in the study of deductive argument. Assumptions are statements that are treated as true for the sake of the argument. It is impractical to make any argument without assumptions. (As an example, there is little point in having to argue that Newton's laws of motion are a suitable approximation for the non-relativistic speeds applicable to AV safety cases. We can assume that Newton's laws apply without a fuss.) But there need to be some ground rules on how assumptions are handled to avoid unreasonable assumptions that compromise safety.

A UL 4600-compatible way of handling assumptions is to use an evidence claim (a claim summarizing a property of a body of evidence) to state the assumption. The evidence can then be whatever seems appropriate for the situation. "Evidence" might simply be an expert opinion, a statement of an industry rule of thumb, or an expedient assumption made to simplify the problem based on personal designer preference.

While an expedient, unsupported assumption might sound at odds with safety, there might be cases where this is reasonable. For example, stating "there will be no kangaroos on the road" might be reasonable for continents other than Australia, and there is likely no point in creating a massive tome of research evidence about this topic. There are bigger safety issues to pursue. (This does not necessarily mean the assumption will be 100% true for a huge fleet of AVs. There might eventually be a zoo escape. Or the use of an AV in a wild animal park's kangaroo enclosure. But it means the design team thinks it is reasonable to make this assumption per the discussion of accepted risks below.)

Other assumptions might be supported by some technical analysis rather than large amounts of empirical data. UL 4600 gives an example of assuming that a green light means cross-traffic has a red light as a reasonable assumption. This might be made based on engineering experience, or might be supported by a reference that traffic light manufacturers in practice follow a standard that supports this assumption. But it should be acceptable to make the assumption that traffic lights in the operational area actually follow such standards rather than having to argue that in detail in the safety case.

There is considerable flexibility as to what type of evidence is needed to support an assumption, so long as the assumption seems reasonable to technically competent designers and assessors. The important thing is that assumptions need to be documented so that they can in fact be checked for reasonableness by someone other than the person creating the safety case. That includes clearly stating the assumption, and being transparent about whatever rationale is being used to justify the assumption. Independent assessment and field engineering feedback are two mechanisms used to help ensure that assumptions actually are reasonable.

5.4.3. Difficult-to-reproduce aspects

An important aspect of an AV safety case is dealing with difficult-to-reproduce aspects of behavior. While exercising the functions of an AV to see if it seems to behave safely on the road is important, any practical amount of testing is not enough on its own to prove safety. (This is true of any complex system using software, but is even more of an issue for systems that use machine learning-based technology and operate on potentially chaotic public roads.)

This issue ripples into many areas of the safety case. For example, evidence based on testing needs to account for statistical significance. Additionally, random variations will need to be inserted to ensure the system is not just getting particularly lucky on selected tests. It also means that potentially elusive software malfunctions such as timing sensitivity (often called race conditions) need to be accounted for.

The standard breaks this topic into two subtly different concepts: nondeterministic and chaotic behavior.

Nondeterministic behavior occurs when a computer program produces different results if run multiple times with exactly the same inputs and starting conditions. This is often because it intentionally uses a randomized process in its computations. Traditionally, safety-critical systems are designed to minimize nondeterminism to make validation easier. However, some techniques, such as AV path planning, can use randomized algorithms.

Even a completely deterministic system will have aspects of non-reproducibility due to a sensitivity to initial conditions beyond the ability of any tester to control. This is chaotic behavior – computational and behavioral conditions are not predictable due to the so-called butterfly effect. Chaotic behavior makes even a deterministic system behave in a nondeterministic way for practical purposes. (Categorizing randomized system behavior for any particular situation as nondeterministic vs. chaotic is not required.)

An example given in the standard for chaotic behavior is that an AV pointed exactly at the center of mass of an obstacle in an unconstrained driving situation might veer either left or right to go around the obstacle. Which way it goes will depend on very slight conditions of internal computational timing, timing of sensor data collection, sensor drift from calibration, vibration of sensor mounts, and so on, which are beyond the practical control of any tester. In practice, there will always be tests for which a single specific response will not be observed every time due to behavioral inflection points. There will always be practical limits to the ability to control a system under test. Any testing must be designed to take this issue into account in general, and recognize that any real-world test results will have some variation due to uncontrollable environmental parameter fluctuations, even if the software itself is written to be perfectly deterministic.

Understanding that systems that interact with the real world via sensors and actuators will not in practice have 100% deterministic behavior is essential for creating a credible safety case. The same system-level test might not have the same behavioral result each time. Moreover, repeated test results might be dramatically different rather than "close" if the test conditions happen to be near a system behavioral inflection point.

5.4.4. Accepted risks

The safety case permits classifying some risks as "accepted," which is different in a very important way from overall risk being "acceptable."

The word "acceptable" in UL 4600 means that whatever is being described will support achieving desired overall safety. So if something is acceptable that means it is expected to make its intended contribution to creating a sound safety case.

On the other hand, an "accepted" risk is one for which a risk – even one with a potentially severe consequence – has not been mitigated as much as it might be (e.g., is not mitigated to as low as reasonably practicable – ALARP), and that lack of mitigation is intentional. In other words, the design team has accepted the burden of the risk with partial or no mitigation. Nonetheless, they deem that the lack of full mitigation will still result in an acceptable item-level risk.

While just accepting some types of risk without justification is likely unreasonable, there will be numerous situations in which an unmitigated or partially mitigated risk is reasonable in the context of a particular safety case. As a hypothetical example, a design team might decide that while the risk of a vehicle being hit by a tornado is real, tornado frequency in the particular geographic area targeted for AV use is low enough that risk will simply be accepted, and the vehicle will not be programmed with tornado detection and evasion logic that might be applied by an experienced human driver. Such an argument might be reasonable where tornados are rare, but not for other areas where they are more common. (Whether such a decision is ethical and/or societally acceptable can be an interesting one, but we do not weigh in on that topic in this book.)

UL 4600 requires that all accepted risks be documented, and that the safety case argues that their contribution to overall risk is small enough that the safety case is overall acceptable. While it might seem obvious that this should be done, this is made explicit in UL 4600 to avoid a situation in which risks that are seen as too small to bother with are rounded down to zero in analysis and then omitted from the safety case as being "unrealistic" or the like. While any individual risk might seem small, in aggregate such accepted risks might total up to a large enough net risk to result in unacceptable safety.

So even seemingly small accepted risks must be traced to the net risk calculation to ensure overall safety is acceptable.

There is an additional requirement to monitor accepted risks via field engineering feedback throughout the lifecycle. This helps ensure that any overly optimistic assumptions are identified and corrected when convenient assumptions regarding putatively negligible risks collide with real-world risk exposure.

Careful consideration of the requirements for accepted risks will indicate that there is very little difference between an accepted risk and a risk for which analysis shows additional mitigation is not needed to achieve the overall required system-level risk. This is the intended goal: ensuring that no risks are swept under the rug, only to come back later and generate surprise safety problems.

5.4.5. Confirmation bias and defeaters

A significant potential pitfall in using a safety case is a tendency to argue that a claim is true without fully considering the reasons it might be false. Confirmation bias is a known issue with any argument process. It might be a particular issue with design engineers who have been through an educational process that tends to focus on how to make things work rather than how things might go wrong.

While there is no perfect way to avoid confirmation bias, the risk of it causing an invalid safety case can be managed in a variety of ways. One is to simply recognize that this is a possibility and consider it when creating and reviewing the safety case. Another is to have people other than developers involved in creating and reviewing the safety case. For example, professional testers often have a somewhat different mindset than design engineers. Design engineers spend their time thinking about how to make something work. Good testers can be ingenious in thinking of ways to make those same things break. Experienced safety engineers can bring to the table lessons learned in other systems of pitfalls and potential failures to avoid. They can also build on numerous prompt elements in UL 4600 that call out specific examples and considerations that should be addressed as applicable.

Having someone with a strong testing or safety mindset try to "break" an argument can help with mitigating confirmation bias. Rather an ask why a claim is true, they can ask why it might be false. In a safety case, this can take the form of a claim being supported by an argument that contains two flavors of sub-claims. One flavor of sub-claim is a supporting reason why the claim is true ("this claim is true because A, B, and C"), as one might expect.

However, a different flavor of sub-claim is that a reason the claim might be false has been considered, and found not to apply. That might sound a little complex, but it is analogous to "we tried a test that might break it, and the system did not break." From an argument point of view, this is a reason why the claim is not false ("this claim might be false because of A or B or C, but all three turn out not to be a problem").

A reason why a claim might be false is called a "defeater" in that it is intended to try to defeat the argument. If a defeater succeeds, the claim is false and something needs to be changed to make the claim true. A defeater that has been considered and turns out to be OK should still be recorded in the safety case to give extra confidence in the comprehensiveness of the argument.

As an example, a claim might be that a vehicle can stop in time if it detects a pedestrian in its line of travel. Sub-claims supporting that claim are likely to be an argument that it can see a pedestrian far enough away that it has time to react and stop before impact. Possible simplified sub-claims: "sees pedestrians reliably at some distance X," and "reaction + stopping time is shorter than distance X." (The value of X is typically a function of speed, but this example is keeping things simple.)

Defeaters would be reasons this might work. For example: "Defeater: road is not too slippery to stop within distance X." (Technically the defeater is "what if the road is too slippery" – but we're phrasing with a "not" so if the sub-claim is true the parent claim is true.) The argument under the defeater would support a conclusion that road friction has been accounted for in the computation of stopping time. If the road is too slippery, the vehicle would either adjust its speed or determine that conditions are unsafe for operation.

5.4.6. Non-deductive arguments and unknowns

In practice, not all defeaters will be considered. Some possible types of factors will be considered outside the reasonable scope of the safety case (which is why fault models are required, as discussed in section 6.1). Other defeaters might be considered so unlikely to be relevant that they are not worth the effort to address in the safety case. And some defeaters might simply be things that the creator and assessors of the safety case did not realize might be relevant.

There are, in principle, an infinite number of defeaters, some of which will be unknown. This is because an AV will need to operate in an unbounded real world which is neither entirely controlled nor entirely understood by system designers – no matter how much they try to control or understand it. So-called black swan events (unknown unknowns) by their very nature cannot be entirely eliminated.

This infinite set of potential defeaters creates a profound shift in how safety cases need to be treated as a logical and philosophical entity. Purely deductive safety cases are typically seen as a starting point when considering safety case approaches. In a deductive safety case there is a conclusion (a claim) that is supported by a string of arguments down to evidence. Even if the argument is based on assumptions, a deductive safety case is considered complete and provably correct within the stated assumptions. In essence, it is a structured mathematical proof. Once the safety case is completed, all claims have been proven true, subject to the correctness of any assumptions.

The scientific method has, over the centuries, come to terms with the fact that purely deductive approaches do not work so nicely in the real world.

That is because of bounds on knowledge and the inevitability of the unexpected. The poster child for this issue is the black swan narrative. (Short version: The thought-to-be-true "fact" that all swans are white was known to Westerners for centuries. Until a European expedition to Australia saw black swans.) This leads to an approach in which claims are not unconditionally true, but rather true as far as is known at a certain point in time. Moreover, a claim has scientific validity only if it is "falsifiable," meaning it can, at least in principle, be proven to be false via some potential real-world observation or experiment. (See: https://en.wikipedia.org/wiki/Falsifiability)

UL 4600 requires treating safety cases as non-deductive. Each claim must be potentially falsifiable in that it must be a statement that could potentially be proven false by some observation. (This property will become essential when we discuss Safety Performance Indicators in chapter 16.) Defeaters must be used to challenge the validity of claims to mitigate confirmation bias. But it is still possible that some relevant defeater will be missing that would have shown the claim to be false, if for no other reason than that the defeater is something that nobody realized was relevant (an unknown unknown).

All this means that a safety case has claims, supporting arguments, and evidence. But, strictly speaking, no claim is unconditionally "true." Rather, it is true as far as is known based on the argument in light of defeaters that have been addressed, evidence available at the time of assessment, and correctness of supporting assumptions. There is always the possibility that some unknown condition will manifest to falsify a safety case claim. The reaction to that needs to be an improvement to the safety case and, in most cases, the system design as well. Such situations are expected in the course of testing and deployment, and are a reason why UL 4600 requires field engineering feedback based on SPIs.

5.5. Safety culture

Another provision in the safety case clause of UL 4600 is a requirement to establish and nurture a strong safety culture. A safety culture reflects the values and practices of how an organization does business with regard to safety.

An informal way of describing a robust safety culture is that: your boss and organization want to hear about safety concerns, and you can count on them to act upon what you tell them without punishing you. In recent years this general idea has crystalized into the notion of "Just Culture," with "just" being used in the sense of "justice," in contrast with a "blame culture" (see: https://en.wikipedia.org/wiki/Just_culture).

Beyond the strictly human cultural aspects, a robust safety culture needs to be supported by processes and activities to ensure that safety is indeed prioritized within the organization. Typically there is a Safety Management System (SMS) for operational safety. Example SMS activities for AVs would include ensuring all testing safety drivers are trained, having viable processes to ground the operational fleet if a significant safety issue is discovered, and keeping metrics on the effectiveness of operational safety practices. The SMS might be extended to other activities such as manufacturing practices. Beyond an SMS, there also needs to be a robust safety engineering activity for the AV design and deployment process.

A robust safety culture is essential for producing a safe AV. If a robust safety culture is indeed in place, it should be easy to provide ample evidence of its existence, including conformance to industry best practices and published guidelines.

5.6. Resources

- ACWG, Assurance Case Guidance, Challenges, Common Issues and Good Practice, SCSC-159, Ver. 1, Aug. 2021. https://scsc.uk/r159:1
- ACWG, Goal Structuring Notation Community Standard. https://scsc.uk/gsn?page=gsn%20standard
- Goodenough, Weinstock & Klein, Eliminative Argumentation: A Basis for Arguing Confidence in System Properties, CMU/SEI-2015-TR-005, 2015. https://www.researchgate.net/publication/272678149_Eliminative_Argumentation_A_Basis_for_Arguing_Confidence_in_System_Properties
- Koopman, P., Kane, A. & Black, J., "Credible Autonomy Safety Arguments," Safety-Critical Systems Symposium, Bristol UK, Feb. 2019. https://users.ece.cmu.edu/~koopman/pubs/Koopman19_SSS_CredibleSafetyArgumentation.pdf
- Stanford Encyclopedia of Philosophy, Defeasible reasoning. https://plato.stanford.edu/entries/reasoning-defeasible
- Wikipedia, Falsifiability. https://en.wikipedia.org/wiki/Falsifiability
- William S. Greenwell, John C. Knight, C. Michael Holloway, and Jacob J. Pease, "A taxonomy of fallacies in system safety arguments," Proc. Int'l System Safety Conference (ISSC), Albuquerque, NM, 2006. https://ntrs.nasa.gov/citations/20060027794

6. Hazards and risks

Summary: Clause 6 of UL 4600 requires addressing hazards, risks, and mitigation in the safety case along the lines typically found in other functional and system safety standards, such as via using safety integrity levels. To retain flexibility it does not prescribe a particular approach, but rather describes the general types of activities that should be addressed in identifying hazards and ensuring risks are mitigated. Risks that can result in harm to people and especially life-critical risks require special attention, going beyond a simple cost/benefit analysis for risk mitigation.

UL 4600 requires arguing that hazards have been identified, risks analyzed, and acceptable mitigation performed to ensure that an acceptable overall system risk is attained. Specific activities that generate work products for the safety case are the identification of fault models, identification of hazards, use of a risk framework, and ensuring that all risks are effectively mitigated.

The terminology associated with hazards and risks is a bit diverse across safety standards. For example, ISO 26262 limits the concept of "harm" to adverse effects on people's health, whereas some other standards also include damage to property and environmental damage. UL 4600 provides flexibility in these regards so long as an overall definition of "safe enough" is stated. UL 4600 uses the term "loss" instead of harm to reduce potential terminology confusion.

Teams should interpret UL 4600 hazard and risk terminology in a way that maintains compatibility with other standards they might be using. Key terms used in UL 4600 for this area are:

- **Hazard:** is not formally defined, providing some flexibility. It is generally intended to mean a potential source or cause of a loss event.
- **Loss:** is defined as a "substantive adverse outcome" with regard to the specific definition of safety relevant to the AV. This is used rather than the term "accident" to avoid any preconceived notion regarding blame, or whether a particular loss event might have been reasonably avoided by a change to a design or operational concept.
- **Risk:** is a combination of the probability and severity of a loss event.
- **Mitigate:** means to ensure that the risk of a loss has been reduced to an acceptable level. Often a hazard is said to be mitigated, but this should be considered a shorthand phrase for mitigating the risk that would otherwise have been associated with that hazard.

The identification of hazards, risk analysis, and risk mitigation are generally compatible with other safety standards, such as the Hazard And Risk Analysis (HARA) requirements of ISO 26262.

6.1. Fault model

A fault model is a representation of the ways in which something can go wrong in a component, system, or technology. This concept heavily overlaps the terms "failure mode" and "failure model," so we will not try to split hairs about the distinctions here. In practice, a fault model might simply be a list of the types of ways a component or function might malfunction that should be considered in the design process.

There might be some ways things can go wrong that are outside the fault model but will nonetheless happen in the real world. Some of those situations might be deemed so improbable as to not be worth engineering analysis (e.g., a direct meteor strike on a moving vehicle). But other omissions might be things that will cause non-zero loss rates in the real world (e.g., lightning strikes on moving vehicles, tornadoes, road flooding, and timing-dependent software defects). Whether any such omission will matter significantly depends on the specifics of the system and its use. Thus, the selection of just the right scope of a fault model is an important system design tradeoff. Things left out of the fault model will not be tallied in predicting expected risk, so the choice of what to leave out must be done with care.

A robust fault model should be used for all relevant aspects of the AV at the system level, component level, and various levels of functionality. This is crucial because hazard analysis (discussed later) uses various types of fault models to determine what hazards need to be mitigated. If a fault model is missing a particular type of fault, that type of fault will go missing from the hazard analysis – but will still find a way to happen in the real world. Worse, if there is no fault model, then engineers will just be doing hazard analysis by the seat-of-the-pants method, with predictably problematic safety outcomes.

Because UL 4600 is a system-level safety standard, the breadth of fault models it requires is significant. While at first the creation of all these fault models might seem like a significant burden, the work would have to be done one way or another. Writing down the list of faults considered in hazard analysis helps defend the thoroughness of the safety engineering, and makes it easier to avoid mistakes of omission in analysis via cross-checking the analysis with the fault model. So creating a robust fault model is best done up front – rather than after a loss event has made it clear something was left out.

The fault models required by UL 4600 are listed below. Numerous specific types of faults must be considered for inclusion in the fault models, with only a few examples listed here:

- **Software:** defects and malfunctions of software including causes such as requirements defects, coding defects, defective exceptional value handling, timing faults
- **Microelectronic and electronic hardware:** power supply faults, single event upset effects, hardware faults, wearout, random failures
- **Sensors:** physical sensor damage, component degradation, loss of calibration

- **Communications:** loss of communication link, packet loss, congestion, undetected loss of data integrity, ineffective communication with humans
- **Data:** data corruption, metadata faults, data ordering faults, data retention length mistakes
- **Electronic/electrical faults:** shorts, cabling faults, water intrusion, thermal, electromagnetic interference, actuator mechanism faults
- **Mechanical operational faults:** vibration, corrosion, freezing, fluid barrier failures, vehicle mechanical faults
- **Procedural:** faults in operational procedure definitions, incorrect maintenance/inspection procedure execution, unauthorized maintenance
- **Item-level:** operation outside ODD, exposure to extreme environmental conditions, invalidation of ODD, intentional misuse/abuse
- **Infrastructure:** missing feature (e.g., no lane marking lines), incorrect feature placement, incorrect feature type, active navigational aid failure.
- **Malicious faults:** cybersecurity and related concerns

Within each fault model, different aspects of faults must also be considered, including:
- Transient vs. permanent faults
- Single faults vs. sufficiently likely multiple concurrent faults
- Common cause multiple faults
- Detected vs. undetected faults
- Accumulation of faults over time

6.2. Hazards

All hazards and risks potentially relevant to safety must be entered into a hazard log. A hazard log tracks all hazards that have been identified during the design and deployment of the system, as well as any mitigation approach used.

While there is, in principle, an unbounded number of potential hazards, the number of hazard log entries is necessarily finite. So there must be some method to determine what to put in the hazard log, and some notion of how many entries is enough. Rather than being a one-off exercise done at one point in the project, the creation of a hazard log should be an activity that proceeds in phases over the development and deployment of the AV.

The hazard log must contain a list of all hazards that have been considered. It is kept updated for the life of the AV, including tracing from each hazard to its resolution (e.g., accepted without mitigation, mitigated, unresolved), an initial risk assigned in the absence of mitigation, a criticality level, and specifically whether the hazard could result in serious injury or death. Risk analysis is discussed in the next section below.

At the beginning of a project, the hazard log can be started by adding lessons learned from previous projects. For autonomous ground vehicles, that can include consideration of hazards and crash types that apply to conventional, human-driven ground vehicles. This does not mean that AVs will be prone to crashing or otherwise failing in exactly the same ways as conventional vehicles. But some amount of overlap in crash types and failures seems likely, and it is a reasonable place to start.

Once an initial hazard log has been created, analysis should be performed that checks what types of faults might present additional hazards to safety. A methodical approach is to check that all types of faults in the relevant fault model have been considered as potential hazards for each component in the system.

A classic technique to discover relevant hazards is Failure Mode and Effects Analysis (FMEA). In simple terms, this involves proposing that a fault from an appropriate fault model is activated in a component, and analyzing whether the results of that fault pose a hazard that is relevant to safety. This technique and its cousins (many of which are listed in UL 4600 section 6.3.1.3.a.) are a "bottom-up" analysis approach in that they hypothesize a fault in some specific component or with some specific aspect of a system. They then determine what type of hazard, if any, that fault exposes the system to. For example, a resistor might burn out (a specific component fault of "fail with an open circuit"), disabling the laser in a lidar sensor, and leading to a loss of distance information to some objects.

Another general type of technique that should also be used is a "top-down" analysis approach, with fault trees being the classical example. In fault trees, a specific loss event or otherwise undesirable outcome is hypothesized, and analysis works backward to determine how that might happen. For example, a loss event might be crashing into an object in the road, which might happen if the AV does not detect the object or the AV is unable to avoid the object. Not detecting an object might be caused by, among other things, a lidar failure, even if it is unclear what might cause that lidar to fail. A top-down approach at some point just assumes some component might fail according to its fault model, without having to be concerned about the physics of how a particular electronic component might malfunction.

Other analysis techniques might be considered non-fault-based, such as including crash scenarios that have happened in the past even if a clear root cause initiating fault was never determined. Those can be book-kept as "hazards" even though they are not associated with any specifically identified component fault (thus the frequent use of the phrase "hazards and risks" in UL 4600 wording).

UL 4600 requires using at least one method of bottom-up analysis, one method of top-down analysis, and one method of non-fault-based analysis. Using more than one of each type of method is a good idea to ensure that the various weaknesses inherent in any specific method are compensated for.

A specific caution is appropriate for hazards created by software. It might be convenient for hazard analysis to make simplifying assumptions about software such as "software failures result in halting execution (a software crash)." While that is certainly one way software can fail, it is by far not the only way. There is always a risk that software will have a design or execution fault that is not just incorrect, but actively harmful (e.g., due to an algorithmic defect that steers the AV into a pedestrian instead of away from one). One hopes that such defects have been eliminated in the design process. However, the only way to credibly mitigate such hazards is to call out an appropriate, documented level of engineering rigor as the mitigation approach. Leaving out the possibility of a complex and potentially harmful software defect during hazard analysis is foolhardy. (Anyone who asserts that software defects always result in simple program crashes is invited to look through the list of automotive software recalls here: https://betterembsw.blogspot.com/p/potentially-deadly-automotive-software.html)

6.3. Risk evaluation

After a hazard has been identified, it is added to the hazard log with an assigned criticality level and an initial risk. The term criticality level is generic to encompass an integrity level or similar concept. For design teams using ISO 26262 concepts, this means assigning an Automotive Safety Integrity Level (ASIL). The criticality level is used to determine a minimum required set of risk mitigation activities.

While selection of a criticality approach is flexible, there is a firm requirement that there be at least two levels of criticality: one for hazards that can plausibly cause fatalities (life-critical risks), and one for hazards that are likely to cause significant human injuries. This implies the existence of at least one more level covering minor or no injuries. Risk evaluation must include harm to both occupants and non-occupants of the AV.

In addition to a criticality level, an initial risk must be assigned. This might be done with a risk table, a risk equation (probability times severity), or other reasonable technique. The risk assigned might be quantitative (e.g., 10^{-4} fatalities per mile from this hazard if not mitigated), or qualitative (e.g., "high" risk according to a defined risk categorization).

There are numerous pitfalls associated with any technique of estimating risk from an unmitigated hazard that are listed in UL 4600 prompt 6.4.1.2.b. There is no perfect approach, so design teams should exercise care in applying analysis techniques depending on their particular system and operational concept. As an example of one pitfall, quantitative techniques are prone to underestimating the risk from high-consequence events that are expected to happen extremely infrequently (one need look no further than the commercial nuclear power industry for cautionary examples – many really big mishaps have this risk analysis issue in common). On the other hand, level-based techniques that do not assume risk can be precisely quantified are prone to systematic under-estimation bias, such as overly optimistic controllability assumptions when assigning ASILs.

There is no requirement to assign a criticality level before vs. after determining the unmitigated risk. But both must be determined and entered into the hazard log to support risk mitigation. That forms the starting point for an ultimate determination of the acceptability of post-mitigation item-level risk.

6.4. Risk mitigation

Once hazards and associated pre-mitigation risks have been identified, the hazard log must also include information about how each risk has been mitigated. Mitigation might be accomplished by removing a hazard, excluding a hazardous situation from the operational design domain, installing technical countermeasures to ensure safety when the hazard manifests, or other approaches.

As an example, a hazard might be identified in that a camera is not able to effectively detect objects in the road when the camera is covered in ice. This hazard might be mitigated by one or more of: excluding cold weather icing conditions from the ODD, coating the camera lens with a hydrophobic layer that prevents water accumulation that might be subject to freezing, adding a sensor cover heater to melt ice, adding a de-icing fluid spray system, adopting a more conservative operational concept if only some cameras are impaired by icing, or other approaches. Some of these approaches might not be effective in all situations, and the mitigation approaches themselves could be subject failing (e.g., running out of de-icing fluid while driving).

6.4.1. Criticality levels and risk mitigation

A requirement of UL 4600 that goes beyond ISO 26262 is that it is not enough to select a particular criticality level on a function-by-function basis and assume the system level will turn out OK. The safety case must also argue that overall item risk is acceptable. This is an extension of typical safety integrity level approaches that includes creating a risk mitigation model (e.g., based on ASILs) and using an identified set of engineering processes and architectural approaches to ensure integrity. (For those using other standards as a baseline, the use of safety integrity levels, performance levels, design assurance levels, and level of rigor approaches are all explicitly permitted as instantiations of the concept of criticality levels.)

For UL 4600, designers must go further and confirm that the integrity-based approach they have used is actually accomplishing the required level of risk mitigation for the overall system. This will include the use of road testing and field engineering feedback data to ensure that mitigations are effective. Ultimately, this motivates the requirement to use Safety Performance Indicators (SPIs) to monitor claims that mitigation of each hazard is performing as required to yield an acceptable post-mitigation risk.

6.4.2. Going beyond risk-based mitigation

Achieving acceptable safety is likely to require more than just satisfying a cost-based risk formula. The ethics of risk management vs. safety are beyond the scope of this book, but suffice it to say that there can be business incentive to deploy a product that is not as safe as many societal stakeholders expect, but that is nonetheless profitable. (If this were not true, the world would not need safety regulators.)

Companies might be especially motivated to deploy not-quite-safe-enough products to meet venture financing performance goals or be first to market. While a safety standard cannot force companies to behave ethically, it can address the difference between risk management and safety by, in part, requiring some safety activities to be performed regardless of perceived cost/benefit within the context of a particularly aggressive business model.

UL 4600 requirement 6.5.2 imposes an obligation to use state-of-the-art practices (as a minimum) when dealing with risks related to potential fatalities and significant injuries to people. This includes adopting relevant safety standards beyond UL 4600 to address such risks. For AVs this is likely to include ISO 26262, ISO 21448, and a number of other emerging industry-issued standards and best practice guidelines. This would in turn impose an obligation to, at a minimum, adopt ASIL-based engineering rigor regardless of whether the design team thought the risk reduction was cost-effective compared to, for example, simply buying insurance that would pay out in case of a fatality (or even just self-insuring fatal loss events by settling court cases).

UL 4600 requirement 6.5.3 imposes a further obligation to avoid common cause and single-point failures for life-critical risks independent of any cost-benefit analysis. This is expressed in terms of analyzing Fault Containment Regions (FCRs).

An FCR is a portion of the design for which any fault can potentially impair all operations for the entire FCR. A typical example is that a single bit flip in a computing unit can potentially cause all the computational results of that computing unit to fail. (For example, what if the bit flip is in a memory location that stores a subroutine return address, and that corrupted return address causes the operating system to crash? Or the program counter itself gets corrupted?)

FCRs extend beyond just computing units to any situation in which common cause failures can occur. So this requirement amounts to saying that any life-critical function cannot have a substantive common cause failure. This extends to also include mitigating any potential accumulation of undetected faults over time that might combine to cause a safety-relevant failure. The FCR phrasing and associated prompts are intended to put the requirement for no single-point failure (and related concepts) on a more rigorous footing than is sometimes seen in conventional automotive systems in the guise of overly-simplistic health checks and sanity checks.

The requirements for paying special attention when risk of harm to people is possible are compatible with ISO 26262, but might at times result in more rigorous engineering requirements than for conventional vehicles. This is appropriate since there is no human driver in an AV to exercise controllability when a malfunction occurs. Alternately, if the automated driving system is tasked with exercising controllability, it must be shown that it will fail independently from and infrequently compared to the hazards it is attempting to mitigate.

6.4.3. Acceptable item-level risk

As mentioned earlier, the item-level risk covered by the entire safety case must be shown to be acceptable. It is not sufficient to simply say that each hazard has been mitigated as much as is reasonably practicable, or that a SIL-based safety standard has been followed. Those techniques might be necessary, but might also be insufficient to achieve overall acceptable safety.

As a specific example, it might be that inclusion of a large number of features at ISO 26262 ASIL D integrity might not be enough in aggregate to reach an acceptable safety target. UL 4600 does not determine how this will turn out, but requires the topic of acceptable overall risk mitigation to be covered by the safety case. Just assuming acceptable system-level safety because some functions were designated ASIL D is not enough.

The requirement to argue item-level risk includes: a hazard log that shows each hazard has been mitigated, traceability from each hazard to supporting argument and evidence of mitigation, and overall acceptability of system-level risk. UL 4600 further requires that even if item-level risk is deemed acceptable at product launch, lifecycle field data must be collected to ensure that the item-level risk remains acceptable via safety performance indicators.

While not explicitly required by UL 4600, it is a great idea to include a complete list of the "safe enough" criteria being used by the design team in the safety case. Some of these criteria will be highly technical, but some will be more social and regulatory matters that are beyond what UL 4600 requires, but still matter for real-world stakeholder acceptance.

6.5. Resources

- ISO 26262:2018 part 3, Concept Phase
- Koopman, P., How Safe Is Safe Enough? Measuring and Predicting Autonomous Vehicle Safety, September 2022.
    - Chapters 3, 4, 5, and 10 are relevant to managing risk and the important differences between risk and safety.
    - Chapter 9 proposes a list of topics that should be included in "safe enough" criteria.
    - See: https://safeautonomy.blogspot.com/2022/09/book-how-safe-is-safe-enough-measuring.html

- Lala & Harper, Architectural principles for safety-critical real-time applications, Proc. IEEE vol. 82 no. 1, Jan 1994, pp. 25-40. https://ieeexplore.ieee.org/document/259424
- Wikipedia, Automotive safety integrity level https://en.wikipedia.org/wiki/Automotive_Safety_Integrity_Level
- Wikipedia, Risk matrix, https://en.wikipedia.org/wiki/Risk_matrix
- Wikipedia, Safety integrity level, https://en.wikipedia.org/wiki/Safety_integrity_level

7. Interaction with people and road users

Summary: Clause 7 of UL 4600 deals with interactions between the AV being described by the safety case, its passengers, and other road users. The clause also covers interactions between the AV and people playing a role in the lifecycle, spanning vehicle transport, operations, and maintenance.

UL 4600 requires arguing that the AV will play well with other road users as well as humans involved in various aspects of operation and the AV lifecycle. Even if an AV is completely autonomous during normal operations, it will continually be in situations in which it needs to interact gracefully with people in a variety of roles.

The need for an autonomous vehicle to interact with people is no different than a human driver's need to interact with other people for a conventional vehicle. That includes any passengers inside the vehicle. It also includes pedestrians, other road users, police directing traffic, and so on. Beyond normal driving behaviors, interactions include other people who need the car to behave a certain way, such as maintenance technicians, tow truck operators, and crash scene responders.

Each type of human actor has behavioral expectations of the AV which are likely to place both obligations and restrictions on safe AV behavior. Some human actors will be benign, but others will range from negligent to impaired to downright malicious. While it might be difficult for an AV to behave with perfect safety in an adversarial road situation, the onus is upon AV designers to anticipate real-world road situations and create an AV that will display reasonable, acceptable behavior.

As you read this chapter you might consider some of the topics covered to be "rare" or beyond what an early deployment of AV technology should have to be able to deal with. That does not change the fact that the topics discussed all happen in the real world, and all contribute in one way or another to the risks dealt with by human drivers. Any particular AV might not be equipped to handle every situation discussed flawlessly. However, the safety case should make clear what is handled by the autonomy system, what is handled in other ways (e.g., via procedures, remote support, or operational restrictions), and whether the result will present acceptable overall risk.

7.1. Communication functions

The safety case must identify and describe all types of equipment and features that are intended to communicate with people, regardless of their road use status and lifecycle role. This is likely to include acoustic devices (chimes, horn, voice), lights, graphical displays, and other input/output modalities. It also includes different types of features for communications input and output such as traffic interaction devices (turn signals, warning flashers), gesture interfaces, occupant distress signals, alarms, microphones, and more.

In addition to overt communication channels, behaviors form implicit communications with other road users. Examples include distinct deceleration to signal intent to stop at a crosswalk, and creeping forward to indicate an intention to move at a multi-way stop intersection when order of precedence might be unclear among multiple stopped vehicles. None of these capabilities are required to be implemented, but they must be identified if used, and accounted for in arguing safety.

UL 4600 describes a wide variety of possible communication channels, and all of them must be disclosed in the safety case to the degree that they might affect operational safety. Other potential communication features include: occupant communication, remote support communication, V2X (vehicle-to-vehicle/infrastructure/other), and communication with people driving other vehicles. Communications can be for a variety of purposes such as communicating motion intent, change of modal state (e.g., entry into autonomous operation), equipment failures, and coordinating passenger ingress/egress.

Any use of communications might help safety by mitigating a risk, or might cause a hazard by either malfunctioning (e.g., telling a passenger to exit a vehicle when it is not safe to do so) or failing when activation is needed for safety (e.g., a warning annunciator that does not activate when required).

7.2. Interacting with people and animals

Regardless of the communication mechanisms used, the AV will need to interact with a variety of people and the occasional animal. Doing so will involve both creating and mitigating a number of hazards. All such interactions need to be identified in the safety case for analysis.

7.2.1. Occupant interactions

Occupant interactions include ensuring safe boarding, such as making sure the vehicle does not start moving while a potential occupant is halfway into the vehicle. They also include ensuring occupants are safe during vehicle operation to, for example, ensure the effectiveness of crash safety devices. A particularly complex area will be ensuring safe occupant egress, including occupants who might legitimately need urgent egress at other than the trip destination, might refuse egress at a dangerous-looking designated designation, or might be too incapacitated to exit the vehicle at the end of a trip (how is a passed out drunk passenger handled?).

Sometimes things other than people will be carried in AVs. Cargo-only considerations that should be addressed in the safety case include proper vehicle loading, securing loads, and ensuring that hazardous cargo is handled safely. Potential misuse must be considered such as people attempting to ride in cargo-only vehicles, whether unattended animals might be transported, and any attempts to put unsafe cargo into an AV such as an open container of gasoline.

A number of other potentially infrequent situations must be considered for inclusion in the safety case if relevant to the safety of off-nominal operational situations. Examples include resolving conflicting instructions from multiple passengers, how to qualify whether a passenger is authorized to override safety features, safety concerns between active missions, movement of auxiliary equipment that might strike passengers or bystanders, potential transportation of maliciously harmful payloads, and so on.

7.2.2. Passenger-settable parameters

A significant area of concern for safety will involve hazards related to passenger-settable operational parameters, if any. What if a passenger is permitted to direct the AV to break the speed limit, override safety commands, or direct the AV to violate traffic laws?

While it might seem odd to permit an AV to break safety rules, there will be scenarios in which a responsible adult passenger will want to reasonably exercise authority to direct the AV to override normal safety protocols, just as a human driver might. As an example, a vehicle might be designed not to drive into a wildfire zone. But a passenger attempting to escape from a town surrounded by wildfires might justifiably see driving down a road with burning brush on both sides as a better choice than remaining in place as their house burns to the ground. Additional scenarios might be directing the AV to ignore traffic signals at obviously empty intersections (with its lights flashing and horn honking) to reach a hospital more quickly during a severe medical emergency, or proceeding through a red light at an intersection instead of waiting for the light to change due to being threatened by a bystander with obvious ill intent.

Violations of safety features might include violating traffic laws, operating outside the ODD, continuing operation despite degraded equipment redundancy, or stopping the vehicle in an otherwise prohibited location.

Such safety override functionality might not be provided by the AV, but if they are the safety case needs to account for them. The safety case should also elaborate on how such situations are (or aren't) handled to provide a comprehensive assessment of overall safety. That assessment will include a more refined definition of ODD that encompasses exceptional situations that the AV is, and is not, designed to handle.

The safety case should explicitly address whether and how the AV handles malicious use and abuse, such as whether passengers are permitted to override normal vehicle behaviors in the face of potential carjackings, assault/robbery directed at occupants, harassment by non-occupants, and how to handle a suspected police impersonation stop attempt.

7.2.3. Demographic profile

People do not all look the same, dress the same, have the same abilities, nor do they act the same. To attain acceptable safety the AV will need to be able to recognize and predict the actions of a wide variety of road users and occupants.

Demographic considerations include variation in how people look: height, shape, weight, posture, age, skin tone, hair style, and clothing. Equipment and baggage people might be carrying or relying upon for mobility will vary: shopping bags, child strollers, canes, walkers, and so on. Abilities will also vary: movement, hearing, vision, speaking, language comprehension, and perhaps temporary impairment of or lapses in judgment.

Special consideration needs to be given to the safety of vulnerable people, including pedestrians and other vulnerable road users in various situations such as: sidewalks, crosswalks, road berms, parking lots, bike lanes, pickup/drop-off zones, vicinity of a school bus, crash scenes, disaster scenes, and so on.

Consideration must be given to situations that arguably are not supposed to happen, but will happen in the real world regardless. Examples include people dashing across the road after debarking a bus (occluded by the bus until they enter the AV's travel lane adjacent to the bus), pedestrians misunderstanding AV signaled intent, or defiant intrusion into an AV's path when the AV should have right of way.

A special concern for safety will be that disadvantaged and minority demographic profiles are not placed at undue risk due to biased training data or other factors. Due to the nature of machine learning-based technology, less frequently encountered types of objects, people, and situations might be handled poorly, so this is a particular concern for AVs.

Risk transfer that puts one demographic at a systematically higher risk than another for reasons of design decisions (or design inadequacy) is likely to create a negative reaction from safety stakeholders. A safety argument that a particular demographic is more difficult to handle for some reason might not be sufficient to resolve such concerns, especially if human drivers would not encounter similar issues.

7.2.4. Safety contribution by people

From a safety point of view, people can help with safety as well as create hazards. Any credit taken for human contributions to safety should be included in the safety case. By the same token, hazards that might be created by people failing to provide expected safety contributions must also be considered. This means that every contribution to safety by people can be counted toward safety, but the possibility of it being omitted or done improperly must also be considered.

Examples of activities for which safety credit might be taken include: pre-departure equipment inspections, passengers following safety rules, other road users following traffic laws (most of the time, anyway), maintenance being done correctly, other vehicles giving way when attempting to merge onto a crowded highway, and so on.

In arguing safety contributions by people, credit can be taken for personnel training, qualification, quality monitoring, statistical analysis of compliance rates, and so on. Potential factors that contribute to non-compliance must be accounted for such as lack of training, honest mistakes, confusion, involvement of children/pets unlikely to comply with behavioral norms, impaired actors, negligence, misuse, and abuse.

7.2.5. Mode changes

Mode confusion is a common theme of many critical system loss events, especially with aircraft. In mode confusion, the system operator has a mental model of the vehicle's operating mode that differs from the system's actual operating mode.

As a simple example, a passenger might think that a vehicle is in fully autonomous mode when in fact it is operating in a driver assistance mode that requires continuous driver safety supervision. That might be because the driver commanded a transition to autonomous mode that failed, the system exited autonomous mode without obtaining confirmation of the change from the driver, or the driver simply lost track of the current automation mode. This can be dangerous if the driver does not realize they are responsible for driving safety while the vehicle is moving.

The safety case must address any situation in which the vehicle transitions to a mode in which greater safety responsibility is placed on a human operator or supervisor. That includes not only driving mode shifts during normal on-road operation, but also special use cases such as vehicle repositioning in a parking area, loading/unloading the vehicle onto a transportation device (e.g., a tow truck), or use of manual controls for maintenance purposes.

In some cases, an unwanted transition into a more highly automated mode can be a hazard, such as when a maintenance procedure is being performed that requires sensors to be activated as if driving is taking place, but the vehicle is actually on a maintenance lift.

The criteria for changing modes must be specified in the safety case along with how risks accompanying such mode changes are mitigated.

7.3. Interacting with other vehicles

For the foreseeable future, AVs will share the road with people in a variety of road user roles, including manually driven vehicles. Moreover, it will be a long time before vehicle-to-vehicle interoperability reaches a state in which safety of movement can be negotiated automatically between AVs, even if AVs happen to be the only relevant road users in a particular situation. Therefore, all AVs on public roads will need the ability to interact with other vehicles that have human drivers (or other vehicles with computers behaving as though they are human drivers).

7.3.1. Hazards involving other vehicles

The safety case must encompass hazards related to interactions with other vehicles. For many AV development teams this argument will involve scenario analysis and simulation results. The safety case should consider other vehicles that are both operated by people and automated drivers, as well as different vehicle types ranging from light mobility devices to heavy trucks to special-purpose vehicles. Scenarios should include not only different driving situations, but also off-nominal behaviors by other road users such as being overly aggressive or excessively cautious compared to other drivers in a particular location.

A different type of hazard involving other vehicles is their emission of energy that might interfere with the AV's own use of the same energy bands. As AVs and active safety systems become more prevalent, the possibility for electromagnetic compatibility issues caused by interference between laser and radar emissions increases and must be accounted for. A related hazard is unintentional interference from other sources including cameras struggling with sun glare at sunrise/sunset.

Finally, unusual vehicle types and road users must be accounted for, including farm equipment, horse-drawn vehicles, farm animals, aircraft sharing the roadway, and exotic vehicle types.

7.3.2. Rule breaking

While other road users are supposed to obey road rules, bending and breaking rules is a regular occurrence on public roads. Some breaking of road rules is due to unintentional mistakes, or even reckless driving. However, some breaking of road rules could be considered justifiable, such as crossing a centerline to go around a downed tree that blocks the travel lane.

One of the reasons that public roads are relatively safe is that human drivers are fairly good at compensating for sloppy driving and driving rule infractions by other drivers. The safety case will need to address defensive driving techniques and safe handling of other-vehicle rule breaking as part of arguing acceptable safety.

While it is popular in regulator-mandated crash reporting and public messaging to blame other drivers for crashes involving an AV, such reasoning carries little weight in a safety argument. A crash is a crash. Blaming other vehicles does not affect the net number of loss events that will be experienced by an AV. Rather, the safety case needs to account for mistakes made by other road users when tallying net overall risk.

7.4. Resources

- Brooks, The Big Problem with Self-Driving Cars Is People, IEEE Spectrum, July 2017.
  https://spectrum.ieee.org/the-big-problem-with-selfdriving-cars-is-people
- IEEE Ethics in action/autonomous and intelligent systems.
  https://ethicsinaction.ieee.org/p7000/
- UK Law Commission, Automated Vehicles Final Report, 2022.
  https://www.lawcom.gov.uk/document/automated-vehicles-final-report/

8. Autonomy functions and support

Summary: Clause 8 of UL 4600 covers the autonomy aspects of the AV, including the autonomy pipeline and associated machine learning-based techniques. Potential hazards specific to each stage of the pipeline are listed for inclusion in the safety case.

The key feature of an AV that makes it autonomous is the so-called autonomy pipeline. This set of functions builds a model of the external world, decides how to proceed within that world, and then commands vehicle motion to perform the driving task. There are a number of significant safety challenges to this arrangement, with the biggest being ensuring the safety of functionality based on machine learning techniques. The standard does not require an AV to depend on aspects of machine learning for safety, but imposes safety case requirements in the common case that the technology is being used.

In several places there are requirements to include elements of engineering analysis in the safety argument that go well beyond a purely black box, statistical simulation/test validation approach to safety.

8.1. Autonomy architecture description

The safety case must include a description of the autonomy architecture including all the parts of the autonomy pipeline (detailed in the below subsections), how they work together, and the redundancy strategy being used. While descriptions of many other electronic features of a vehicle might be fragmented into a set of feature descriptions, the autonomy pipeline must work as an integrated whole, so there is a requirement for a comprehensive overview of its functions in one place in the safety case.

Even though the set of functions is typically called a "pipeline," data does not necessarily flow in strictly one direction. Additionally, functions might be divided differently by a design team compared to the categories listed. So the UL 4600 requirement is that prompts be covered by the safety case, not that they be organized in precisely the same way as the listed functions.

The engineering process of a system that incorporates machine learning-based technology is quite data intensive, which makes it different than embedded development processes that are focused on software source code. Because machine learning involves training based on massive amounts of data, the description of the autonomy architecture must include a description of the data management practices being used, including data selection, data cleaning, algorithm selection, execution architecture selection, and the training approach.

A redundancy management strategy is required, but might be tricky to get just right. Many AVs will have a primary autonomy channel and a backup autonomy capability. Dependably switching from the primary to the backup will require an accurate way to know that the primary channel has in fact failed, which might be especially difficult for perception failures. If the backup channel is a less capable mechanism for executing a safing mission, the primary channel will need to be careful not to put the AV into a situation that is beyond the capability of the backup channel to recover. Any commonality between a primary and backup channel in sensor use, algorithms, or possibly even training data used for perception algorithms might be the basis of a common cause failure that undermines the dependability of the backup channel. In practice, arguing that multiple different autonomy capabilities within the same AV will fail independently is challenging.

The requirement for an architectural description is intended to result in a theory of operation manual for the autonomous aspects of the item as well as the engineering methods used to create functionality based on machine learning technology. The importance of this architectural description is that it provides a framework for understanding and assessing hazards that affect how different parts of the pipeline might work together safely – or fail to accomplish that.

8.2. Operational Design Domain

The Operational Design Domain (ODD) is an abstraction – some would say a model – of the circumstances in which the AV is designed to operate. The safety case must argue that the AV will be acceptably safe when operating within the ODD. For that to result in real-world safety, there are a number of factors that must be addressed, including completeness of the ODD description and what happens when the real world does not match the ODD.

The ODD must cover all relevant aspects of the operational environment. That includes not only geographic location, but also road surface characteristics, bridges, local driver behavioral patterns, weather extremes, types of missions being executed by the AV, vulnerable road users present, general traffic rules, special road user rules, seasonal effects, and so on.

The completeness criterion for the ODD description is a bit subtle. What matters are not the factors that a human driver would think are relevant to environmental and other conditions. Rather, factors relevant to the safety of the automated driving system controlling the AV are what matter for the ODD. Many factors will be similar. But some factors that would not matter to a human driver – such as whether trees have leaves on them, changing their lidar reflectivity – might matter greatly to an AV, and therefore would need to be accounted for in the ODD. The important point is that the ODD need not be a model of the world relevant to a human driver, but rather must be a model of the world that is relevant to the particular automated driver that is the subject of the safety case. The relevant factors to include in an ODD will not be quite what a human driver would consider important, and can be expected to vary across different AV designs.

Handling an ODD violation (in which the AV operates outside the ODD) requires special care and attention. Because the ODD is much more than a geographic limitation, there will be cases in which it is not possible to have a timely warning that the AV is about to exit the ODD. A sudden rain squall, a cargo truck spill into the path of the AV, a bridge collapse, a flash flood, or other event can forcibly eject the AV from its ODD while it is in motion, with insufficient warning to complete a safety shutdown before exiting the ODD. This means the safety case must argue that safety outside the ODD is acceptable given the operational concept. This is likely to involve an approach in which the AV performs a safety stop as quickly as possible to minimize its exposure to the unknowns present when operating outside its ODD.

The issue of potential ejection from the ODD places an additional constraint on ODD selection. If an ODD and accompanying system design are so brittle that they are continually violated by real-world operational conditions, overall system safety might be unacceptable even if, strictly speaking, safety might be OK within the ODD. Imagine a potential ODD that excludes all pedestrians and has no way to detect them. A vehicle safe within that no-pedestrians ODD might have safety issues when encountering a person repairing a flat tire on a limited access highway due to the ODD (no people are supposed to be present) failing to match the real-world presence of a person.

No matter how much design effort is expended, having real-world events eject the AV from its ODD is always possible. In principle, any time the AV encounters an object type or traffic situation relevant to safety that it has not been trained on, it has been ejected from its ODD. Therefore, the safety case must argue that detection and response to ODD departures is acceptably safe. Data regarding ODD departures must be fed back to the design process and tracked to resolution.

The risk of ODD ejections might be mitigated by having a very broad region of the ODD that is relevant only to operation during fallback maneuvers (executing a safing behavior to terminate a mission). This aspect of the safety case is up to the design team, but merits careful thought to minimize the frequency of ODD ejections and ensure net overall acceptable safety.

The ODD should be expected to change over time. In some cases, the ODD will need to change to track changes in the real world to avoid an elevated ODD ejection rate. In other cases, the AV design will be updated to change or expand its operational abilities, introducing novelty that needs to be added to the ODD. Either way, the safety case must be updated to reflect any changes to the ODD.

There are somewhat open terminology issues regarding what exactly an ODD might be, whether there are multiple ODDs for a particular vehicle, whether there might be many different sub-types of ODDs, and so on. Eventually, standards will evolve to address these topics, but for now our position on ODDs (which is in alignment with UL 4600) is the following:

- **There is exactly one ODD** for an AV's autonomous operation mode. The safety case is tied directly to that ODD. Changing the ODD requires changing the safety case to match. In principle a vehicle might have multiple ODDs, for example associated with different deployment scenarios or business models. But each ODD would need its own separate safety case, and only one ODD can be active at any given time for any specific vehicle.
- **ODD "subsets" can be a useful** concept. For example, one ODD subset might be operating during the day, and another operating in a rainy night. The AV might use different sensors and control algorithms for different ODD subsets. The safety case might argue various aspects of safety differently for different ODD subsets, but ultimately the whole ODD must be accounted for in the safety case, with no gaps left uncovered by the union of all ODD subsets.
- **The term ODD should not be used to refer to the real world.** The ODD is a model, which necessarily is an abstraction of the thing it is modeling (the real world). As the saying goes, the map is not the territory. We prefer the term "Operational Domain" for the real world itself vs. ODD for the engineering model of the real world. However, UL 4600 does not designate a specific term for the real world in this sense.

8.3. Sensing

The safety case must argue that sensors and sensor fusion work acceptably well. There is some complexity here because what is acceptable for any particular sensor depends upon the needs of the perception system, the availability of redundant and diverse sensor types, the operational concept, and so on. A high-speed truck operating on limited-access highways will have different sensor needs compared to a low-speed shuttle operating in congested urban centers. The argument for sensor adequacy will need to take all these factors into account.

For the purposes of the current version of UL 4600, high definition maps are considered a synthetic sensor that can be used in conjunction with other environmental sensors. Localization is considered a sensor function, but is not treated in depth beyond a requirement for localization fault management. In practice, performing a safing maneuver without crashing after loss of localization information is a somewhat complex but important issue to address in the safety case.

8.3.1. Individual sensors

No sensor can perfectly sense the external world. There will always be noise in sensor data, as well as environmental conditions that impair sensing abilities. The safety case should take these issues into account when making claims about sensor capabilities regarding correctness, completeness, and timeliness of data being provided to the autonomy pipeline. Sensor data will not be perfect, but it only needs to be good enough for the perception function to work well enough to assure acceptable safety.

A key requirement for sensors will be ensuring proper calibration. Calibration should address differences among individual components during initial AV manufacturing and repair part installation. It will also need to address component aging, degradation over time (e.g., lens scratches), physical mounting changes (e.g., bent mounting brackets), and needs for routine maintenance such as cleaning.

The safety of sensors that emit energy must be considered. While some sensor emissions are rated as safe regardless of use, others might have operational conditions imposed such as requiring continuous movement of a laser beam to avoid damage from a fixed beam position entering someone's eye for a prolonged time, or disabling radar transmissions when people are likely to be close enough to the emitter to risk ocular damage, depending on emitter frequency and radiated energy.

8.3.2. Sensor fusion

An AV is likely to have multiple sensors to ensure dependability in the face of component failures as well as to provide diverse data sources. The sensor fusion capability combines those multiple data sources to provide higher quality data regarding the external world than would be possible with a single data source.

The safety case must describe the sensor redundancy strategy and redundancy management plan, arguing that net sensing performance across the sensor suite is acceptable. For example, if a safety case uses ODD subsets, this is likely to involve arguing that the particular sensor fusion strategy used for each ODD subset is acceptable for that subset's sensing conditions. If the overall claim structure is that radar and lidar will work when cameras are ineffective, then there will need to be two different arguments: one for the camera case, and a different one for the radar+lidar case, with each linked to a different ODD subset.

Sensor redundancy management can be complex. Different sensors might disagree as to whether an object is present. The argument must be careful not to contradict itself about the use of redundancy. For example, consider an argument that says that at least one of three types of sensors will see a pedestrian, and a different part of the argument that says that if two out of three sensors agree, the third disagreeing sensor will be outvoted. If only one sensor sees a pedestrian does that count as a detection (at least one out of three sees the pedestrian)? Or is that sensor out-voted as a false detection (two out of three do not see the pedestrian, out-voting the one that sees them)? Any inconsistencies of this type need to be identified when assessing the safety case.

Any credit taken for sensor diversity must be justified with quantitative data. It is not sufficient to simply assert that sensors are somehow "diverse" and therefore will have zero failure correlation. The effects of the diversity to cover gaps in other sensor type performance and failures must be worked out in detail. Especially important is considering potential common cause sensor failures due to sensing gaps, adverse environmental conditions, or zonal failures (e.g., road debris damages multiple different types of sensors mounted in a single sensor pod).

8.4. Perception

Perception is the process of taking sensor inputs and transforming them into a model of the external world. That model will not be perfect due to the limits of sensors and the limits of techniques used to infer characteristics of the real world. (Safety considerations specific to machine learning are covered in the next section.)

The safety case must argue that perception has acceptable latency and performance metrics. Key metrics called out for consideration are false negatives (e.g., did not see a pedestrian that is in fact there) and false positives (e.g., AV does a panic stop due to sensing a phantom pedestrian that is not really there).

An important consideration is that blanket perception statistics are unlikely to tell the full story about safety in uncommon circumstances. For example, certain types of road users might be mischaracterized at a very high rate even though the system works well on average. A system that correctly perceives pedestrians 99% of the time might still fail to detect any construction workers in high-viz vests at all as pedestrians. If construction workers are less than 1% of all pedestrians, the system might really be 99%+ accurate on everyone except those wearing high-viz vests, with a 0% detection ability for construction workers in high-viz clothing.

Perception performance must be argued to provide acceptable coverage of a perception ontology, and must map sensor inputs to that ontology. The ontology includes both objects and events, with sensor inputs processed by the perception system to match up to categories in the ontology. This, in effect, forces some amount of interpretability onto the machine learning process. The perception system must map real-world objects and events onto a set of ontology elements that makes sense to a human designer.

The argument in support of perception must also justify classification error rates, biases, and training data coverage of the perception ontology. Coverage of perception validation must be traced to the ODD and any subsets to ensure that all relevant objects, events, and driving situations are represented suitably within the world model being built by the perception system.

8.5. Machine learning

Validating the use of machine learning-based techniques for life-critical systems is no easy task, and UL 4600 does not have a silver bullet solution to that challenge. However, since the goal of the standard is to ensure that safety cases address potential issues, it does call out a number of considerations that must be addressed to support any claim of acceptable safety.

The safety case must describe how machine learning (and any other artificial intelligence technique) is used in the system, including both how the machine learning is accomplished and how the results of that machine learning are deployed in the AV. An argument that the results are suitable for use must cover both the implementation (is the execution engine that runs a neural network programmed correctly?) as well as the performance (did the system learn what it needed to learn, and does it perform well at the system level?).

Machine learning practices that might be thought of as good hygiene must be addressed in the safety case. This includes avoiding overfitting, ensuring that validation data information has not leaked into training data, considering the effects of data labeling inaccuracies, avoiding data bias, and ensuring coverage of data across the entire ODD.

The integrity of data used for training and validation must be addressed, since safety outcomes depend so heavily upon that data for machine learning techniques. This includes the integrity of the data collection system, the data storage system, and data management systems that perform training and validation. Regardless of the capabilities of machine learning techniques being used, if they are created and validated using biased or corrupted data, the results are unlikely to be safe in practice.

A known potential issue in this area is sensitivity to data variations. Small, essentially imperceptible, changes in data values can cause significant behavioral changes in a machine learning-based system. Hazards due to data variation must be accounted for due to sensor noise, environmental variation, and operation beyond environmental parameters associated with the machine learning system's design and training.

Changes to machine learning data that involve retraining must be handled carefully to avoid compromising safety. Even adding a single new training data element and retraining a system might produce safety-relevant changes in system behavior, complicating impact analysis in the face of changes. Online learning during vehicle operation that changes behavior outside the engineering validation process must be handled with extreme care, and is likely to be unacceptable if it can result in potentially unsafe behavior.

8.6. Planning

The safety case must describe the strategy and algorithms used for planning vehicle motion. That includes the approach for ensuring obstacle avoidance as well as determining safe and feasible control actions. Plans created by the system must be acceptable both in terms of safety and also in terms of being executable by the control system.

Avoiding collisions involves setting obstacle clearance buffers, which might be in terms of distance, time separation, or some combination. Buffers are a matter of both engineering margin and practical safety in case loose straps or the gesturing arm of a pedestrian get caught on a vehicle protrusion that passes too closely.

While collisions and near hits are undesirable, they are sure to occur in the real world eventually. The planner must include considerations for how to avoid high risk situations proactively, and what to do if a crash is inevitable. There is no requirement to solve the infamous Trolley Problem (how to select victims in a no-win situation – see Section 10.5). Rather, there is a requirement to describe what approach is taken to collision-imminent situations as part of an argument justifying acceptable overall safety.

Specific attention must be paid to the validation challenges involved in nondeterministic planner behavior. Eventually a planner will fail during operation by, for example, failing to find a safe plan quickly enough to meet the required timeline. The safety case must explain how such failures will be handled, potentially involving transitioning the AV to a safing trajectory to terminate the mission.

One aspect of planning that can be complex is how the AV's behavior might shift risk onto other road participants. For example, an AV that performs an in-lane stop in the middle of a high-speed motorway might not be blamed directly for any crashes, but could easily be at elevated risk of being hit from behind by a large truck.

8.7. Prediction

While prediction is sometimes not called out as a separate autonomy pipeline stage, this function is present one way or another in any autonomous system, perhaps spread across perception and planning functions. It is treated as a separate function here in terms of the safety case because of its key importance.

To avoid collisions, an AV must not only know where various objects are at the moment, but also whether they are going to overlap with its planned position anytime in the future. An ability to predict the possibility of future collisions requires estimating the future motions of other objects. The degree of sophistication required for prediction will depend on the ODD and the AV's driving aggressiveness. There will also be situations in which the AV's

Autonomy functions and support

motion affects the plans of other vehicles, which are likely to change their motions in response.

For some situations involving slow movement and relatively slow changes in behavior, a simple extrapolation of current movement might be sufficient. However, in some circumstances objects might change their motion quite dramatically in short periods of time, such as a pedestrian changing from standing still to running across an intersection to catch a bus that has just arrived at the other side of the street. Those changes must also be predicted to avoid some types of collisions.

There will be situations in which worst case predictions make it impossible for the AV to move at all. It is acceptable to argue a probabilistic approach (most pedestrians will not suddenly run across the road mid-block without warning). However, any predictions must be validated at the time of release, and monitored for continuing accuracy via field engineering feedback (in some contexts a pedestrian running out into the street might be quite likely, even if that context is somewhat rare on average). As with other aspects of safety, blaming other road users for unpredictable movement will not change the fact that a crash has occurred due to a prediction failure. The question for each such crash is how the system might be improved to reduce the risk of future crashes.

8.8. Trajectory, motion control, actuation, and status

Once a path has been planned based on a world model that includes prediction of other actor motions, the AV must determine its own motion and execute a planned trajectory. That trajectory might be a series of waypoints, or a series of velocities and accelerations. Regardless, the vehicle needs to be commanded to move including both speed changes and direction changes.

Ensuring safety will require faithfully executing the trajectory intended by the planner. However, it will also require respecting the operational limits of the vehicle in its current conditions. For example, vehicles will have turning radius limits to avoid rollover at high speeds, and limits on ability to brake based on vehicle load, equipment status, and road conditions. The planner should not be counting on a vehicle to perform beyond its physical limits. If the vehicle determines it unexpectedly cannot fulfill a planner command, there must be a fault handling process in place to determine how to resolve the situation.

The interface between the automation system and the underlying vehicle controls must resolve potential safety issues having to do with vehicle equipment, especially if an autonomy kit has been added to a conventional vehicle. Vehicle status warnings and alarms displayed on a dashboard for action by a human driver will need to be interpreted and acted upon by the AV. Potentially some actions such as repairs will involve intervention by human mechanics, but the safety case must still show that any status alerts will be acted upon as required, rather than noted without action because the AV is incapable of responding appropriately.

Another issue for an autonomy kit is that the AV controls might operate the vehicle in a way that is atypical of human drivers, and that falls outside the validation performed for the underlying vehicle platform. As a hypothetical example, pressing a physical accelerator pedal takes a certain amount of time for a human driver. An automated driving system imposing an instantaneous voltage change representing an instantaneous large change in acceleration command might cause problems, especially if acceleration commands cycle rapidly for some reason. (For example, a pulse-width modulated signal instead of a smoothed analog signal might in principle cause a dramatic vehicle malfunction, but only in some very specific situations.)

Similarly, an instantaneous rather than smooth steering angle command change might destabilize a steer-by-wire directional control loop. The safety case must argue that any validation of the underlying vehicle platform is predictive of safe operation – despite any differences in commands sent by the automated driver compared to a human driver.

Finally, the automated driving system must be able to handle vehicle failures that are not annunciated by vehicle monitoring systems. Wheels falling off, brakes failing, suspension failures, steering mechanisms falling apart, high voltage propulsion batteries catching fire, and so on all happen on real vehicles. The automation must be able to compensate for such failures at least to the degree that controllability credit was taken for a human driver in the safety assessment of the vehicle platform.

8.9. Timing

Since the autonomy pipeline controls vehicle motion in response to external events, timing considerations are emphasized in this aspect of the safety case. Timing requirements include ensuring that end-to-end timing including computing latency through the autonomy pipeline from sensor to actuator is fast enough to control the vehicle and react to any driving hazards.

Timing must be acceptable not only in normal situations, but also in worst case situations that could occur within the ODD. This is likely to require graceful degradation techniques, especially for perception. If the number of objects sensed is too many to handle with available computing resources, the AV perception system cannot simply slow down to the point that it no longer meets real time deadlines. The highest-risk objects are likely to be prioritized, or the vehicle might need to slow down its speed to give it more time to react.

Loss of control loop stability is another concern, involving a computational lag large enough that control corrections come too late to keep the vehicle on its intended trajectory. Control loop stability analysis should be included in the safety case. Consideration should be given to the effect of

Autonomy functions and support   67

any communication lag for any control activities performed off-vehicle on the far end of a communications link.

8.10. Resources

- Data Safety Initiative Working Group (DSIWG), Data Safety Guidance v.3.4, SCSC-127G, Feb. 2019. https://scsc.uk/scsc-127G
- ISO 21448:2022, Road vehicles – Safety of the intended function.
- Koopman, Short course on AV safety. https://users.ece.cmu.edu/~koopman/lectures/index.html#av
- Koopman, P. & Fratrik, F., "How many operational design domains, objects, and events?" SafeAI 2019, AAAI, Jan 27, 2019. https://users.ece.cmu.edu/~koopman/pubs/Koopman19_SAFE_AI_ODD_OEDR.pdf
- Safety First for Automated Driving, 2019. https://group.mercedes-benz.com/dokumente/innovation/sonstiges/safety-first-for-automated-driving.pdf

9. Software & system engineering process

Summary: Clause 9 of UL 4600 requires suitable software and system engineering process rigor to ensure safety. This includes following best practice overall engineering processes, ensuring high software quality, and ensuring high software development process quality.

While data engineering and machine learning quality matter, there will still be plenty of conventional software involved with an AV (including the execution engine for machine learning-based algorithms). This clause requires conforming to best practices for software and system engineering quality.

9.1. Software and system engineering key points

The practices described in UL 4600 are generally representative of software quality practices required by other safety standards such as ISO 26262. These processes are ubiquitous across safety-critical industries for conventional software engineering (or should be), so we will not belabor the point here.

Here is a summary of key software and system engineering points required of the safety case, and in particular, arguments that software and system engineering processes have contributed to safety:

- Follows best practices for software and system engineering that are traceable to industry standards and professional society-issued guidelines. These must include: effective peer reviews, software testing, conformance to a coding style standard, and static analysis.
- All process steps produce written, auditable work products.
- Engineering processes are defined and mappable to encompass all the development steps in a functional or system safety standard (e.g., IEC 61508, ISO 26262, MIL-STD-882). Any process approach, including Agile, may be used, so long as it somehow includes comparable process activities and work products to ensure appropriate safety integrity.
- Addresses configuration management, version control, quality assurance, and process quality assurance. Requires defined software quality acceptance criteria beyond test results (e.g., evidence of effective peer reviews).
- Defines and validates safety requirements.
- Addresses safety for hardware, software, third-party components, and includes autonomy components in the scope of quality/safety engineering activities to the degree they are relevant to safety.

- Collects and analyzes defect data for both software defects and process defects.
- Records and justifies deviations from processes.

9.2. Resources

- ANSI EIA-649-C 2019 Configuration Management Standard
- Boehm et al., Balancing Agility and Discipline: A Guide for the Perplexed, 1st Edition, 2003
- Koopman, Better Embedded System Software, 2022 (revised edition). ISBN-13: 979-8596008050
- Koopman, 18-642 Course Lectures, Carnegie Mellon University. https://users.ece.cmu.edu/~koopman/lectures/index.html#642
- SEBOK: https://www.sebokwiki.org
- SWEBOK: https://www.computer.org/web/swebok
- SEI, "+SAFE, V1.2: A Safety Extension to CMMI-DEV, V1.2," https://resources.sei.cmu.edu/library/asset-view.cfm?assetid=8219
- Wikipedia, Configuration management. https://en.wikipedia.org/wiki/Configuration_management

10. Dependability

Summary: Clause 10 of UL 4600 addresses dependability, including degraded operating modes, redundancy management, fault detection, robustness, post-crash behaviors, system timing, and cybersecurity.

The term dependability is used as an umbrella term to encompass fault tolerance, robustness, and other properties that have to do with the AV acting in a safe way despite something going wrong, ranging from a flaw in the design to an unavoidable crash.

10.1. Fault detection and mitigation

All equipment operating in the real world will break one way or another. A central theme in dependability is having a layered approach to avoid faults, tolerate faults, and prevent faults from escalating into more serious system issues. The first line of defense is removing systematic faults at design time via high integrity design practices and safety engineering. But the next line of defense is fault detection and mitigation during operation.

Any type of safety-critical equipment needs the ability to detect its own faults and respond to mitigate the effect of those faults. The types of faults that must be detected and reacted to are defined by the relevant fault models required by UL 4600 clause 6.

10.1.1. Fault detection coverage & diagnosis

Detecting faults is easier said than done. It is not enough to do some cursory health checks and declare that fault detection has been checked off in the design process. As a simple example, using the watchdog timer built into a microcontroller is an excellent idea (if implemented properly), but is far from a comprehensive fault detection mechanism.

Coverage against the stated fault model is a central concept for fault detection. Fault detection coverage matters because components can have subtle, dangerous ways of failing. Sometimes a faulty computer just stops working. But sometimes it generates incorrect, dangerous answers. Moreover, planning to switch to a redundant backup component does not help if the system does not detect a component failure to initiate the switching process. And what if the switching mechanism is faulty, or takes so long to switch that a crash happens while the system is between modes?

Worse, what if a switching mechanism activates a backup component that is itself faulty? The safety case will need to argue that fault coverage is high enough that the probability of a component with undetected faults being called upon to ensure safety during operation is acceptably low.

Achieving high fault detection coverage generally requires a multi-prong approach. Some faults can be detected while the vehicle is in normal operation. Those are the easy ones. If a task stops responding, a data value fails sanity checks, or an error detection code triggers when retrieving a memory value, it is clear that something has gone wrong. Additionally, if redundant computations that are supposed to return identical answers return different answers, the system can tell that at least one of the redundant computations is incorrect.

With a little more effort, some faults can be actively detected during operation, such as a memory scrubbing task that reads memory locations – even if otherwise unused – to check for memory error coding violations.

Other faults will be difficult to test during operation, and require offline tests between missions or during maintenance procedures. As a simple example, testing that a washing fluid reservoir "empty" sensor works properly might require emptying the reservoir on purpose to see if the sensor triggers as a periodic maintenance procedure.

Some reliability math will be needed to make sure that offline tests are conducted often enough to avoid long exposures to undetected faults or the possibility of multiple redundant components all becoming faulty between checks. In some cases there will be a tradeoff between how often components are tested vs. how many redundant components need to be installed to ensure redundancy is not depleted by accumulated component failures between checks. In vehicles that might operate continuously for most of a day or even multiple days, this might require different diagnosis strategies than current vehicles that do many of their tests at the start or end of each ignition cycle.

10.1.2. Reintegration

The problem of reintegration must also be considered. After a fault has been detected, the faulty component might be taken offline. But how is a previously faulty component brought back into service?

A component that fails in service might be disabled during a mission. Is it brought back into service during the mission? Or is it brought back into service at the start of the next mission? How does the system know that it has actually been repaired between missions?

An underlying fault might not be detected at system startup because it is intermittent, or because it only occurs in certain operational conditions that are not present during start-of-mission testing. So there might be no way for the system to distinguish between a previously faulty component that has been repaired and one that simply has faults not detectable by self-tests. Automatically returning a previously failed component to service because it passes a cursory self-test during system startup can expose the system to unacceptable risk if it fails again during the next mission. Especially problematic is the potential for multiple copies of a component to accumulate matching failures that are not detected during self-test.

Serious consideration needs to be given to how and when to bring a previously failed component back into service. Highly critical component

Dependability

failures that might be missed by self-tests should not be automatically cleared, but rather require a physical maintenance activity to ensure the component has been replaced or thoroughly subjected to an offline test.

Even less critical component failures should be logged and analyzed to detect broader problems beyond a single one-off transient glitch (for example, caused by a radiation-induced single event upset). For example, repeated patterns of intermittent failures in a particular component should trigger a maintenance activity even if any individual failure was not high risk.

10.2. Redundancy management

Components will eventually fail, and a redundancy management plan must be in place to respond to those failures. The safety case must list the different types of redundancy present and their role in supporting dependability. That includes the physical architecture (computing hardware, sensors, actuators), the software architecture (different functions), and the mapping from software to hardware.

The point of having redundancy is to ensure that when one component fails there is a defined response. Unlike typical functions in a conventional vehicle, an AV must fail operational in that the computer must continue to exercise control over the vehicle and ensure safety despite a component failure. There is no human driver to compensate for equipment failures. (For this discussion we consider a software function to be a "component" regardless of the details of its implementation.)

The basic unit of redundancy management is the Fault Containment Region (FCR; see Section 6.4.2). For redundancy management purposes an FCR is a combination of hardware and software that is either working or failed as a single unit. The concept of "partially working" does not apply, because once there is a fault inside an FCR, no guarantees can be made about the rest of the FCR. On the other hand, the failure of one FCR should not induce failures in any other properly designed FCR. (If that is not true, then the thing that failed is by definition not an entire FCR).

Redundancy only counts if there are no common cause failures between components that are intended to be FCRs. This means not only using different computing resources, but also requires careful analysis to avoid depending upon possible compromise of shared or co-located resources such as: power supplies, cooling systems, sensors, actuators, wire harness connectors, network connections, tool chains, operating systems, software libraries, and redundancy management software. In cases in which there are components that are shared or have potential common cause failures (e.g., two copies of the same software potentially sharing a design defect) then the safety case must argue that design integrity is high enough that any systematic failure is unlikely to happen due to the design and manufacturing process being a suitably close approximation of perfect. (This can be done, but doing so is quite challenging.)

Safety-critical architectures commonly use an isolation strategy to avoid failures in one FCR propagating into another FCR. This might include physical isolation of memory storage devices, physical isolation of computing units, hardware separation of task address spaces, and so on. Soft isolation such as using memory management units to isolate address spaces in memory might suffice, but the possibility of that isolation mechanism forming a source of common cause failure must be considered.

Ultimately, the point of having redundancy is usually to accomplish some combination of two purposes:
- **Fault detection.** A second FCR is used to detect a flawed result or behavior by a primary FCR via comparing outputs, monitoring liveness, or computing some sort of acceptance check.
- **Improved availability.** A second FCR is used to take over if a primary FCR fails, for example to provide a fail-operational instead of fail-stop capability. Additional redundancy might be added to further improve availability.

The purpose of any redundancy must be explained to ensure that redundancy use is not improperly double-counted in the safety argument. For example if two FCRs are used to cross-check an answer (fault detection), the argument cannot properly claim that the same two FCRs provide a primary/backup failover pair. (If they disagree, how would the system know which one is correct?) There might be subtle redundancy arguments to make, but clearly identifying the purpose for each redundant unit in the safety case can help expose such potential over-commitment of redundancy claims.

As an example of explaining redundancy management, consider a triplex modular redundant system that has three channels that "vote" on an answer. If two out of three agree, the third is ignored. Out of the three channels, one is there for basic functionality (if you only had one, you would get an answer, but not know if is right or wrong). The second is there for fault detection (if the two disagree you know there is a fault, but not which channel to believe). The third is there for improved availability (if any pair agrees you can ignore the faulty third channel).

10.3. Operational modes

Every system has different operational modes. If nothing else, there is an "on" mode and an "off" mode. However, any vehicle has a number of additional modes that go beyond the on/off mode selection, especially for automated equipment. A special subset of those modes that requires additional analysis is degraded operations that occur because of equipment or functionality failures.

10.3.1. Identifying modes

The safety case must identify all modes that the system can be in. These often involve different hazards, different behavioral requirements, and ultimately differing safety arguments. Some example modes and sample considerations are below:

- **Normal operation mode:** normal driving
- **Emergency vehicle mode:** make way for emergency vehicles
- **Parked:** power off moving mechanical elements; avoid roll-away incidents
- **Refuel/recharge:** immobilize vehicle to avoid fueling hose/recharge cable and infrastructure damage
- **Post-incident:** immobilize vehicle to avoid injury to victims and responders; unlock/open access doors for rescue; de-energize high voltage power cables
- **Maintenance:** energize designated equipment for maintenance and diagnostics without (necessarily) enabling other vehicle movement

It is possible that some operational modes can be collapsed into a smaller number of vehicle modes. Nonetheless, the safety case must argue that all potentially different operational situations beyond normal driving have been accounted for.

A related issue is that safety must be ensured when changing operational modes. For example, transitioning from refuel/recharge to normal operation requires removing the connection to the fuel hose or charging cable. (Can the AV be sure this has happened? Or does some human maintenance operator confirm to the AV that it is free of entanglements?) Changing from maintenance mode to normal operational mode requires ensuring that all maintenance has been completed and the vehicle is no longer in a special maintenance condition.

10.3.2. Degraded operations

A crucial part of safety engineering is ensuring that the system will remain safe even after something goes wrong. At some point, built-in redundancy will be depleted by accumulated component failures. Moreover, even if functionality can be sustained, depletion of any redundancy is likely to alter mission parameters to deal with the risk of catastrophic losses if the one remaining operational component fails before repairs can be made.

A key concept for degraded operations is having a Minimum Equipment List (MEL). The MEL is a list of all components that must be non-faulted, including redundant components, to operate safely in a particular operational mode. This list might well contain more components than are required simply for functionality, because it includes any redundancy required to meet safety targets accounting for mid-mission failures.

As an example, the MEL for a two-engine aircraft departing a coast on a trans-Atlantic flight is two engines – even if it can safely fly on only one engine. To be a bit simplistic, after takeoff the second engine is not primarily needed for thrust – it is there to have redundancy in case the first engine fails mid-flight. If one engine has failed since takeoff, the aircraft degrades to a diversion operational mode in which it lands at a nearby airport rather than continuing its flight. The engine failure rates and probability math work out the maximum distance the aircraft is allowed to fly from a diversion airport to ensure that it always has a place to land sufficiently close in case of an engine failure. If both engines fail in flight the aircraft does not meet its MEL for powered flight and the pilots do the best they can. (See https://en.wikipedia.org/wiki/Gimli_Glider for an out-of-fuel condition and https://en.wikipedia.org/wiki/US_Airways_Flight_1549 for common cause engine damage). Some older aircraft designs had three engines, in large part so that two would be left operable in case of a mid-ocean engine failure.

As with aircraft, if an AV has two autonomy computers in case one fails, it is important that both computers be operational at the start of a mission. In other words, the MEL for starting a mission would be both computers operational. If one of the computers fails during operation, the mission might need to be cut short, with a MEL of only one computer supporting a degraded operational mode with a very short maximum remaining mission length. And it is always conceivable that both computers will fail during a mission, so there should be some best effort plan to deal with a situation in which available equipment does meet any degraded mode's MEL for continued operation.

The safety case for an AV must list all degraded mode configurations and argue acceptable safety for each of them. This includes the safety strategy for each degraded mode, what subset of devices and computation functions must be working for that mode to be active, and why that is acceptably safe within the context of the system-level operational concept and risk budget.

There is also a requirement to argue acceptable safety for a catastrophic equipment failure in which no defined driving mode MEL requirement can be satisfied. That does not mean that a catastrophic failure must be free of the risk of a loss event, but rather that the net risk is acceptably low. The net risk reckoning can take into account how infrequently catastrophic failures will happen and the fraction of time any last-ditch safety mechanisms such as uncontrolled panic braking are expected to work out favorably.

As a matter of practicality, the degraded modes are likely to have fairly big "steps" in terms of MEL differences to reduce the number of different configurations that must be validated and addressed by the safety case.

A sketch of a hypothetical arrangement that might be included in more detail in a safety case could be the following modes:

- **Fully operational:** everything is working. Normal missions.
- **Degraded redundancy:** full functionality, but installed standby spare components have been configured to stand in for failed components. A maintenance session has been scheduled for the next off-peak maintenance availability. The probability of a worse failure mode has increased, but is judged acceptable given remaining redundancy.

- **Finish mission:** equipment that is necessary to meet dependability requirements for a full day of use has failed, but all functions are still operable. The current mission will be completed, and the vehicle will then proceed to the maintenance facility before the end of its usual daily operational cycle. The probability of a worse failure mode has increased, but is judged acceptable given remaining redundancy to complete a short mission.
- **Terminate mission:** a failure precludes continuing the mission. The vehicle proceeds to a safe termination location within an operational limit. (Similar to an aircraft landing at the nearest viable airport.) The operational limit is in place to manage the risk of further equipment failure. This is likely to also involve less aggressive driving, such as an AV version of reducing speed to slower than other traffic, turning on 4-way flashers, and looking for a place to pull onto the berm. There are various ways to handle this situation depending on the specifics of the failure and how much effort the design team has put into defining and validating different approaches to handling this condition gracefully.
- **Emergency controlled stop:** a significant failure prevents a graceful mission termination. The vehicle does the best it can given the circumstances, potentially resulting in a controlled in-lane stop.
- **Panic stop:** a catastrophic failure has impaired vehicle controllability. An emergency brake is activated that stops the vehicle as fast as possible with best effort control (for example, steering angle locked in current position and brakes applied for a maximum-deceleration stop).

The safety argument for this arrangement would describe the modes and MEL requirements for each mode, with "panic stop" being the default if no other MEL is met. It would then argue that each successively worse degraded mode is successively less likely to occur. Moreover, each mode is designed to either obtain maintenance or terminate the mission with high probability before a further degradation forces downgrading to the MEL of an even more degraded mode. The safety risk of each mode takes into account both the risk of operating in that mode (in-lane stop is risky in high-speed traffic) as well as the probability of being in that mode (most of the time the vehicle will be able to pull over to the berm, exit the highway, or be in a situation with only low-speed traffic; in-lane stop and panic braking should be very rare).

A key consideration for designing degraded operational modes is that if a single function or component is present in all MELs, if that single point of failure happens to fail, it will drop the system all the way to a panic stop mode. This especially applies to sensors that are shared between MELs, but also sensors shared between different redundant computation channels.

An additional consideration is that for degraded operational modes to work, the system has to recognize that a component has actually failed. Some failures are obvious (e.g., a software crash). But other failures might be difficult to detect, increasing the risk of operating unsafely rather than trigging a switch to a degraded mode. For example, if a perception function starts having a high false negative rate due to operating in an unusual environment, degradation to a "go slow and avoid any object regardless of classification" mode can't happen until the system realizes that its perception performance has been compromised.

Some modes might be fully supported by a fairly modest MEL. For example, automatic parking might be supported even after several sensors for high speed driving have failed. Thought should be given to how graceful degradation relates to operational mode shifts.

Additionally, safety during mode changes can be tricky to handle, but must be considered. As an example, if equipment failure reduces the maximum safe operational speed for the vehicle, how is safety ensured during the time it might take to slow the vehicle down to the new speed limit?

Once a degraded mode has been entered, the system should not spontaneously elevate itself into a more capable mode without attention from a qualified maintenance activity. This is to avoid mode thrashing in the face of an intermittent component fault, and reduce risk during the window between the time the fault might reoccur, the system detects it, and the system can stabilize operation in a degraded mode after each triggering of the intermittent fault.

10.4. Item robustness

No engineered product is perfect, and no engineered system perfectly accounts for what can happen in the real world. The AV will need to be robust in the face of faults, operating conditions, environments, and behaviors that do not go as expected. (Some people use the term resilience here in addition to or instead of robustness. We will not split definitional hairs here.)

The AV needs to provide acceptable safety even when things go wrong. That is true even if (especially if) the things that go wrong were unforeseen by the design team. This does not mean that the AV has to provide full functionality, but rather that the system cannot be so brittle that it operates dangerously at the first whiff of a surprise. A few areas that need to be considered for robustness include:

- Being forcibly ejected from the ODD by outside forces not considered during the design phase
- The AV not behaving as commanded, possibly due to an equipment fault or environmental condition not considered during design
- Finding out the hard way (during operation) that an assumption made in the safety case is false
- Detecting that the system's expectations have been violated, such as an object moving in an unpredicted manner
- Other road users behaving in ways that are illegal, unexpected, or previously thought to have been impossible

In simple terms, the world is full of surprises. People, despite their many limitations, are impressively good at knowing a surprise has occurred and trying to do something reasonable in response. Machine learning-based systems, on the other hand, are notoriously brittle when presented with situations that are different in some important way from their training data. Despite this challenge, AVs must be reasonably robust in the face of the unexpected to ensure safety.

10.5. Incident response

Regardless of how well designed and operated an AV is, eventually it will be involved in a crash or other incident. While the safety case might argue that the AV will cause no at-fault crashes, loss events will still find a way to occur, and the AV will need to respond appropriately.

Incident response involves a number of design considerations that go well beyond the behaviors of normal driving. Drivers are generally required by driving laws to stop after a potential loss event and potentially render aid. Additionally, if there is potential damage to the vehicle or risk to passengers, the AV needs to facilitate passenger evacuation and ensure crash scene safety to the degree it is able.

First, and perhaps most challenging, the vehicle needs to actually know that it has been a participant in a loss event. For severe crashes, this should be straightforward. If the airbags deploy, it is pretty obvious there has been an incident. However, other situations are more subtle. A gentle bump (from a multi-ton vehicle perspective) could cause significant harm to a vulnerable road user through impact, or even a glancing contact such as a sideswipe. Scenarios such as brushing against a bicyclist that causes them to crash, or snagging loose clothing on a pedestrian dragging them down the street might happen. The design team must foresee different types of incidents other than major crashes and make sure that the AV can determine something problematic has happened and respond appropriately.

Even before a potential loss event occurs, the AV should be doing what it can to reduce potential severity (or, if possible, improve the odds that a loss event will be avoided). A key strategy in most situations is reducing vehicle speed. Sometimes a maneuver might be undertaken to attempt to dodge a possible collision. There is risk here because the system has imperfect knowledge of the outside world, so it is possible a maneuver made to avoid one bad outcome might unwittingly cause a different bad outcome by hitting something undetected.

This area of discussion tends to invoke the much-debated "Trolley Problem" (see: https://en.wikipedia.org/wiki/Trolley_problem). UL 4600 does not require design teams to solve that philosophical problem. Rather, the AV needs to have a defined, reasonable set of requirements or rules for how it is expected to react to detected impending loss events, and rules for what it can and cannot do in response to such events.

As a simplistic example, the AV's rules might be that it maneuvers to avoid a crash if doing so avoids crashes with all other road users with high probability, but if not, it uses maximum possible deceleration to minimize impact speed. It might also have a policy that collision into fixed obstacles that are not road users (trees, etc.) is preferable to collision into vulnerable road users due to protection afforded occupants by passive safety equipment. However, UL 4600 does not require these specific rules. Rather, it requires that these types of considerations be addressed by the safety case.

An incident might result in a loss event, or might be a near-hit situation in which the vehicle is unequipped to evaluate the severity of any loss that might have occurred. Regardless, the vehicle needs to react in an appropriate manner. Typically that will involve coming to a stop and calling a human operational supervisor for help, as well as deploying whatever local features it has that are responsive to the situation.

It is likely that the AV will need to enter a special post-incident mode to ensure that it does not display dangerous behavior at the crash scene. The AV design team should consider functionality that needs to be activated and deactivated post-incident, especially in situations involving equipment damage to the AV. One example is occupant evacuation, such as for a battery fire or other urgent hazard to occupants. Other examples include deactivating moving equipment that might cause a hazard after a crash, de-energizing high voltage electrical systems, and activating other safing functions as appropriate. Different types of incidents might require different responses.

The AV will need to communicate and cooperate with emergency responders at a crash scene. That includes letting them know vehicle status (for example, that motion in fact has been disabled). It also includes providing relevant behaviors and changes to equipment status as might be done by a human driver, ranging from moving a vehicle out of the flow of traffic despite significant vehicle damage (if the vehicle is capable of moving) to configuring the vehicle for towing.

If it is appropriate to return the AV to service after an incident, it is important to ensure that any potential equipment damage has been accounted for and any legal requirements to remain at an incident scene have been resolved.

After any incident, there should be an analysis to ensure that correct behaviors were displayed, and to determine if any safety case improvement or design change might help reduce risk from future incidents involving similar factors.

Data recording needs to be done with acceptable forensic validity. A specific concern here is that a computer system that causes an incident cannot be trusted to report reliably on factors that led to the incident without independent checks. For example, a speed controller that reports it was commanded to operate at maximum torque might have been commanded to do so, but the fault might be that it interpreted a torque command incorrectly. Data being reported by a component that contributed to an incident cannot be taken at face value, because a fault in that component might also result in incorrect data reporting. Put another way, if you are attempting to rule out a

Dependability

fault in a component as a contributor to a loss event, you should not place unquestioning trust in data reported by that same potentially malfunctioning component.

10.6. System timing

Vehicle operation requires close attention to real time control and real time task scheduling. The real world does not wait for computations to be completed as the AV is hurtling down a highway. While section 8.9 covers the timing of autonomy functions, this section covers timing of all the other parts of the system.

Using a Real Time Operating System (RTOS) to manage the timing of safety-critical automotive functions is well-plowed ground. To be sure, doing so requires effort and attention to detail. In particular, newer computing architectures can make it difficult to determine the Worst Case Execution Time (WCET) for each task, as is required by scheduling methodologies such as Rate Monotonic Scheduling (RMS), Deadline Monotonic Scheduling (DMS), or Time Triggered (TT) scheduling. Nonetheless, timing of conventional software can be managed using state of the art automotive techniques.

The difficult part for an AV will be creating a framework in which the vehicle control RTOS can co-exist with the potentially much different timing behavior of the autonomy functions. The WCET of a perception system might be difficult to bound in any useful way, because it could depend on the number of objects or the classification of objects as it affects the complexity of prediction algorithms. Nonetheless, there must be a way to manage real time properties for the system as a whole in the face of finite computational power available to model an open-ended real world and respond appropriately. Likely this will involve some degraded functionality scheme as discussed in section 8.9.

There is a requirement to detect and mitigate any violation of real time requirements or resource exhaustion. This is to ensure that if computations take too long compared to how fast the real world is changing, the AV will be able to detect it has a problem that needs to be dealt with to ensure safety.

10.7. Cybersecurity

An AV that does not have adequate cybersecurity protection measures in place is unlikely to be safe. An adversary might modify program images, manipulate data, or alter the environment in a way that causes an incident. UL 4600 is not a security standard, so this aspect of safety is limited to a single clause. Nonetheless, addressing cybersecurity issues is essential for safety.

The safety case must reference a cybersecurity plan that addresses different aspects of security including confidentiality, integrity, and availability. Cybersecurity issues that are relevant to safety should be exported from the cybersecurity plan into the safety case so that they can be taken into account in understanding the full safety picture. This includes identifying malicious (adversarial) faults in relevant fault models.

As a simple example, a threat to software integrity can be a toolchain attack in which a subverted compiler injects malicious code into the compiled executable while doing compilation. This type of hazard should be included in the safety-relevant fault model for software, but mitigation might be a reference to the portion of the cybersecurity plan that deals with tool chain integrity, potentially coordinating with other tool chain qualification activities.

While many of the faults relevant to cybersecurity are classic computer security issues, some topics dealt with in this section bridge the gap between cybersecurity and physical attacks, as is common in security considerations for machine learning-based perception systems. For example, alteration of street signs or inflicting physical damage on sensors might impair the ability of the system to operate properly in ways not readily detected by the system itself. While the AV cannot prevent malicious actors from defacing street signs, the fault model needs to address the possibility that this can happen in a way that fools perception algorithms, while not necessarily being evident to human drivers.

10.8. Resources

- Avizienis, A; Laprie, J.-C.; Randell, B.; Landwehr, C. "Basic concepts and taxonomy of dependable and secure computing," IEEE Transactions on Dependable and Secure Computing, Jan.-March 2004, Volume: 1 Issue:1 pp.: 11-33. https://doi.org/10.1109/TDSC.2004.2
- ISO/SAE 21434, Road Vehicles – Cybersecurity Engineering, 2021.
- Knight, John., Fundamentals of Dependable Computing for Software Engineers, 2017, ISBN 1138402222.
- Koopman, Safety Architecture Patterns: https://youtu.be/QEHr8J-ByLQ
- Wendover Productions, Small Planes Over Big Oceans (ETOPS Explained): https://youtu.be/HSxSgbNQi-g

11. Data and networking

Summary: Clause 11 of UL 4600 deals with ensuring the integrity of data, spanning data transmission, data storage, and various uses of data to support the AV. It also covers various types of infrastructure, both data and otherwise. Given the emphasis on field engineering feedback in UL 4600, many types of data have increased safety significance compared to conventional vehicles.

The traditional emphasis on safety-critical system design has been on functionality and ensuring high quality software source code. However, the importance of data integrity and the role of data in determining safe functionality dramatically increases when using a machine learning-based approach that is validated largely via simulation.

The safety case must address data communications, data storage, and infrastructure support. Remote operation is also addressed by this clause. Cybersecurity is relevant to the topic of data safety, but is discussed in section 10.7.

11.1. Why data safety matters

An important consideration for data safety is the overall increase in the significance of data to the design process for an AV compared to a conventional vehicle. In a conventional vehicle, for example built according to ISO 26262, it is common for only some data in the vehicle (e.g., the program image, calibration data, and some control loop data) to be considered safety critical. If the vehicle passes testing, and the diagnostics give adequate coverage of equipment failures, those two aspects might be considered good enough for safety, and the rest of the data is treated as much less critical.

However, in an AV there is a much broader scope of data safety that must be considered. Data that becomes much more critical for ensuring safety needs to be addressed by the safety case, potentially including:

- **Training/validation data.** If the safety case argues that the AV's machine learning has been trained and validated on data that represents the ODD, then that data is relevant to safety. Bias, corruption, or gaps in the training and validation data will mean that the machine learning does not cover the ODD as thoroughly as has been argued in the safety case.
- **Simulation data.** Consider if the safety case argues that simulation shows safety, and road testing is just there for a sanity check on the simulation. That means data used to build the simulated world, simulated actors, and simulated own-vehicle behaviors all become safety critical. Also relevant to safety are data concerning the orchestration and

reporting of simulation campaigns. For example, if the pass/fail data log for simulation is inaccurate, the AV might be reported as safe per simulation results even if it failed some automated simulation tests – but those failures were not properly recorded in simulation result logs.
- **Safety Performance Indicator data.** If the safety case argues that safety will be acceptable in part because any newly emergent problems will be detected and corrected via SPIs before a loss event is likely to occur, then the entire SPI data flow from sensor to SPI dashboards is relevant to safety. If an SPI violation is not reported, that undermines the safety case.
- **Map data.** If the safety case argues that high definition mapping data will be accurate and updated quickly to reflect changes in the environment, then map data becomes safety critical.
- **Environmental data feeds.** Real time weather data feeds, infrastructure feeds to vehicles (e.g., from traffic signals), and the like all become safety critical to the degree that the safety case argues that they mitigate risk.

The degree to which engineering and design data become safety critical depends on the specific processes and technology being used. However, data safety is becoming increasingly relevant for a wide variety of conventional embedded systems, and the increased emphasis on using data in the creation of AVs makes this even more of a concern for AV safety cases.

A related issue is that even some current safety-critical system designs do not fully account for the safety implications of using low-integrity sensor data in high-integrity functions. The safety case is required to enumerate all sources and uses of data that are relevant to safety-critical functionality. This is intended to call attention to improper uses of low integrity data within the system.

11.2. Data transmission

Data flowing within a component, among components, between a vehicle and its support infrastructure, and between a vehicle and other communication partners outside the scope of the item must be considered in constructing the safety case. This includes not only computer networks (wired, wireless), but also human interfaces, service tools, video capture, text messages, and more indirect data transfer methods such as scanning QR codes.

11.2.1. Data transfer properties

The safety case must identify all data transfer means that might be relevant to safety, which goes well beyond the usual computer networking connections. Transfer methods include in-vehicle WiFi, Bluetooth, data access ports (e.g. USB interfaces), remote keyless entry channels, analog inputs, diagnostic ports, data connectivity to infotainment devices, map data, plug-in temporary driving controls, data logging, navigational positioning signals, vehicle-to-infrastructure communications, and even electrical switches. Any way that data can be transferred in or out of the system or among components might affect safety.

Each data flow must be analyzed for relevance to safety, and hazards associated with each data flow as appropriate. Properties often relevant to safety include data value accuracy, authenticity of data source, timeliness, and availability (lack of data transfer outages).

A subtle but crucial issue in data transmission has to do with data flows across functions with different integrity levels in a system. If a highly safety-critical function needs data input, it cannot necessarily trust data being provided by a lower integrity function, even if that data originally came from a high integrity data source. For example, a speed sensor might have no redundancy and be prone to faults that output incorrect speed values with no indication that a failure is taking place. Or a low-integrity electronic control unit might read data from a high integrity speed sensor and forward it to a high criticality component. A life-critical function that needs high integrity speed values should not simply accept the output of a low integrity component without further fault mitigation approaches. In general, a computation that is assigned to a particular integrity level should ensure that all its data inputs have at least as high a data integrity level. While this might sound obvious when stated this way, lapses in this principle can be found in real system designs.

In practice, the issue of ensuring sufficient data integrity can propagate into many far recesses of a system. As an example, if the AV depends on the correctness of map data for safety, that propagates a requirement for high integrity out to the data communication network used to provide map data, the cloud server repository used as a source of map data, map data handling software, sources of information used to create the map, and so on.

Ultimately, all data used in high integrity computations must be traced back to their source to ensure suitable integrity. In some cases, low integrity segments of the data flow can be tolerated by adding high integrity data wrappers so long as the data is being forwarded unchanged, but the case must be made that end-to-end integrity has been assured.

11.2.2. Remote operation

Teleoperation is a special type of data flow that requires particular scrutiny. The role the remote operator plays in safety must be identified. This might range from direct teleoperated driving, to remote safety supervision, to remote takeover after automation failure.

Issues of particular concern include authentication (is the remote driver actually authorized to operate the vehicle?), what happens when remote control connectivity is lost during operation, what happens when data rates are reduced while the AV is in motion, and how variable or excessive communication latency is handled.

Special caution is required for potential infrastructure failures. While a vehicle might have redundant communications via two different mobile data service providers, those two providers might fail concurrently due to shared cell tower infrastructure, shared backhaul data conduits, signal interference from geographic features such as tunnels, local interference from unlicensed transmitters, and so on. It can be expected that there will be wireless communication link outages, and the AV should be designed to be safe when that occurs during operation.

UL 4600 does not encompass the specifics of human/machine interface issues for remote driving. For example, it does not set requirements for minimum control loop latency to ensure good control characteristics, nor requirements for displays to ensure any teleoperator has sufficient situational awareness to operate or supervise an AV safely. These are important issues that should be addressed in the safety case, but fall in an area that was put outside the scope of the standard.

11.3. Data storage

The storage and retrieval of data bring additional safety concerns. The safety case must identify the different types of data and storage mechanisms used for safety-relevant data. That includes data beyond such obvious things as program executable images and on-vehicle configuration data. Any data that can compromise safety must be identified and addressed by the safety case.

On-vehicle data is perhaps the most obvious, including executable program images, configuration parameters (typically called calibration data in the automotive industry), neural network configuration for any machine learning-based components, data logs, and metadata such as the current version of a data set. This includes not only data for the primary vehicle controller, but also safety-relevant data in sensors, actuators, and any other electronic control units that could be relevant to safety.

Managing remote support data is also important. This especially includes any data hosted in a data center that supports the AV. Some of this data might be relevant to real-time operation such as temporary map updates (e.g., active construction site locations), weather advisories relevant to ODD limitations, and so on.

Other data might be more durable, but hosted off-vehicle nonetheless to be pulled in on demand. This might include high definition map updates, software version updates, and on-demand support materials for maintenance, inspection, and occupant informational videos.

A third type of data that must be managed is engineering lifecycle data. This includes SPI-related data collected from both the vehicle and lifecycle processes that is processed and stored for engineering feedback. Engineering feedback data is used to improve the system, lifecycle processes, and the safety case. Also relevant are data logs associated with recording faults, failures, incidents, and mishaps, especially to the degree that historical data from root cause analysis is used to support the safety case.

The fault model considered for data storage needs to be thorough. Data can be corrupted via storage media failure. But it can also suffer other fates, such as: corruption of metadata, loss of removable media devices, inadvertent deletion of data during data storage maintenance, illicit deletion of data that might make a stakeholder look bad, bulk loss of data due to disasters, loss of data due to over-write when local storage capacity is exceeded, and loss of index information that permits locating specific data. Data can also be corrupted indirectly via changing data formats, changing measurement units, and reordering of ordered data if metadata is not in place to mitigate such issues.

Data might retain its integrity while still being inaccessible due to an unavailability of software tools to ingest the data, unavailability of hardware to access a data storage device, and format drift that makes older versions of data records non-importable to newer processing tools.

11.4. Infrastructure (data and otherwise)

Section 11.4 of UL 4600 covers the area of infrastructure support. Some aspects cover computer-readable data, while others are more about physical infrastructure. However, the boundaries can be difficult to separate for a design approach that uses simulation heavily, so both aspects are included in one place. For example, if a roadway intersection changes, that is a physical change to the real world, but has important implications for map data and possibly whether the intersection is no longer within the ODD. Similarly, "smart infrastructure" might provide data to an AV, but also relates to physical objects and roadway features in the real world. And even "dumb infrastructure" such as painted road lines provide coded data to AV sensors (line color and whether lane markers are dashed, solid, or doubled encodes roadway information for the AV to read using sensors).

The safety case must include safety-relevant aspects of infrastructure that are necessary for safety-critical operation. This ranges from road construction (painted stripes, road geometric limits, pavement surface characteristics), to navigational support (GNSS navigational support, beacons), road signage, and special signaling conventions.

All aspects of the environment that the vehicle depends on as part of its ODD need to be included in the safety case as infrastructure assumptions. One way to approach the topic is to consider what environmental aspects need to be included in a simulation model to fully span the ODD. This includes not only road types and object types, but also signaling conventions ranging from AV-specific transponders to human-readable traffic signals to manual traffic direction gestures from emergency responders.

Hazards associated with various infrastructure aspects need to be identified and mitigated, including whether the vehicle is able to handle the presence, absence, or malfunction of infrastructure features safely.

11.5. Resources

- Driscoll, K., Hall, B., Koopman, P., Ray, J., DeWalt, M., Data Network Evaluation Criteria Handbook, AR-09/24, FAA, 2009. https://users.ece.cmu.edu/~koopman/pubs/faa09-24_data_network_evaluation_criteria_handbook.pdf
- SCSC Group: Data Safety Initiative: https://data-safety.tech/
    - DSI Data Safety Guidance, with the latest version being SCSC-127G version 3.4: https://scsc.uk/scsc-127G

12. Verification, validation, and test

Summary: Clause 12 of UL 4600 deals with verification, validation, and test approaches that must be encompassed by the safety case. The related topics of run-time monitoring and safety case updates are also included.

While testing alone cannot prove that a life-critical system is safe, testing is certainly an important part of a safety-critical engineering process. Beyond testing, the safety case also includes other verification and validation techniques that assure quality, and serve to provide checks and balances to the design process as well as lifecycle safety performance. Run-time safety monitoring provides a run-time mechanism to detect safety issues during testing and deployed operation. The result of issues detected by these various techniques inform changes to the system as well as updates to the safety case.

12.1. V&V methods

The safety case must identify the various Verification and Validation (V&V) methods used as part of the design process and lifecycle. The usual distinction between the two activities is that verification is ensuring that the results of one step of the design process are properly incorporated into the next step of the design process. Validation, on the other hand, is checking that the outcome of a design step meets associated requirements. Informally, it can be said that verification is "did we build the thing we intended to build?" while validation is "does the thing we built do what it needs to do?" While there are important philosophical differences between verification and validation, UL 4600 does not require any fine distinction to be made as to whether any particular activity falls into the verification vs. validation category. Rather, such activities are lumped as "V&V" and the design team is left to sort out the division of activities according to what works best for them and any other safety standards they are using.

Types of V&V methods that should be used include: peer reviews, static analysis (compiler warnings and beyond), formal proofs, unit testing, integration testing, system testing, robustness testing, stress testing, and fault injection. Different approaches to testing should include testing of isolated software components, software-in-the-loop testing (e.g., simulations), hardware-in-the-loop testing (e.g., test benches), and vehicle level testing.

Just as important is tying the V&V method being used to its contribution to the safety case via a traceability strategy. If you ran a test, what property was the test supposed to demonstrate that relates to safety? If you formally proved a property of a system, how is that property relevant to safety? Ultimately the point of doing V&V activities is to supply evidence in support of a specific safety claim. That might be that a particular fault mitigation works properly, or it might be something more indirect such as a claim of engineering rigor having produced a suitably low design defect rate.

Key observations are: (1) there is more to V&V than vehicle-level testing, (2) there is more to V&V than just testing, and (3) V&V activities are only productive if they provide evidence to support the safety case. Other activities unrelated to the safety case might be useful for requirements discovery and the like, but are not really V&V.

12.2. V&V coverage

V&V coverage refers to the degree to which V&V activities exercise potential faults to see if they can be detected in multiple dimensions spanning not just technology, but also design and lifecycle phases. Ultimately, the safety case will need to rely on claims that the design is correct and complete. There will also need to be claims that each vehicle is built correctly and will not encounter safety problems due to run-time faults. V&V coverage is the degree to which all of those claims are supported by V&V evidence.

Some aspects of coverage amount to checking whether V&V aligns with the different faults considered within the fault model (see clause 6) for each component. However, V&V coverage goes beyond just fault model coverage, because not every problem stems directly from a specific type of fault in a particular component. For example, two components might each be fault free according to their respective specifications, but be combined in a way that does not provide acceptably safe vehicle behavior.

Mandatory for inclusion in V&V coverage analysis are: systematic design faults, data faults, requirements problems, violations of requirement assumptions, a clear identification of the ODD accompanied by full ODD coverage, and an analysis that each type of hazard has been acceptably mitigated. Hazards go beyond simple design and operation faults, including maintenance faults, operational procedural definition faults, problems with external data sources, and potential issues with third party components. It does not matter whether something is under the control of the AV design team or not. If a fault could cause a safety problem, then V&V must ensure proper mitigation.

12.3. Testing

While testing is a specific V&V method, it has unique requirements that deserve special attention. As a starting point, performing testing requires a test plan (what behavior or other aspect of the system is being tested?), exercising some aspect of the system (actually running the test including setting up initial conditions and executing a specific piece of functionality), a test oracle (what is the correct result from performing the test?), testing coverage analysis (how thorough was the test per some relevant metric?), and traceability of the test results to the safety argument. If it is missing any of those aspects, it is not really testing.

It is common to hear that some activity is "testing" when it is really just someone messing around with the system. Consider taking an AV out for a 100 mile road trip on public roads (hopefully with a competent safety driver). To actually be testing, that exercise would have to include: a test plan (what specific behaviors were being checked on those 100 miles, and were the 100 miles planned to exercise behaviors in need of validation?), the actual driving part (were test conditions achieved to exercise intended functionality?), a test oracle (what did designers expect to happen for each segment of the drive, and how did that compare to what actually happened?), test coverage analysis (how much of the system was exercised by the test drive?), and traceability of the test results to the safety argument (which claims of the safety case does the test coverage attained support?).

A simple statement such as "we drove 100 miles on public roads without crashing" involves an implicit test plan of driving some non-specific 100 miles (were they "easy" miles, in "easy" weather?), driving, a test oracle of "did not crash," and support to the safety argument of 100 miles of crash-free driving against a target of hundreds of millions of miles. Such a test might feel like progress, but in reality, contributes little to the safety case. In other words, such an exercise is less about testing and more about messing around from a safety case point of view.

This is not to say that opportunistic road testing has no benefit, but rather that a lot more work needs to be undertaken to turn such an exercise into relevant testing. A common approach is to back-fit road experience to create a retrospective testing plan. Resimulation (using road data to drive a simulation) might need to be done to infer an after-the-fact test plan, determine what parts of the system were exercised, compare actual vehicle behavior against simulated behavioral predictions (amounting to an oracle), and then add the resulting test data to evidence supporting whatever claims happen to be relevant to the experience.

While gathering road testing evidence provides significant early evidentiary support for the safety case, that works best to cover the most common test cases. Over time, more care must be taken to fill in coverage gaps. Various rare road situations, infrequent operational scenarios, unusual road objects, extreme weather conditions, equipment failures, and so on will need to be sought out on public roads or tested via simulation. Supporting the safety case not only requires enough testing evidence in terms of bulk, but also enough of all the various different kinds of test evidence needed to span V&V coverage requirements.

If a test fails (the test result does not match the oracle prediction), the issue might be a defect in the AV or its design. Or the problem might be a defect in the test plan, an incorrect oracle prediction, a test execution error, or other issue with the test itself. Regardless of the source, every test failure needs to be investigated, the problem corrected, and revisited to ensure that all tests required for safety case coverage pass.

Two specific types of tests beyond ordinary exercise of functionality are required. Regression tests are needed to validate any changes, especially after the AV has been deployed. Additionally, fault injection testing is required to ensure that any credit claimed for fault mitigation is appropriate. If a claim is made that the effects of a fault will be mitigated, then some sort of validation is required to determine that the mitigation approach will actually be effective.

12.4. Run-time monitoring

Run-time monitoring can be a close cousin of testing. The idea with run-time monitoring is to instrument the AV so that if something goes wrong, that situation triggers some sort of alert, warning, safety response, or other appropriate reaction. It is like testing in that an oracle (in the form of a run-time monitor) determines if observed behavior is correct. However, unlike testing, a run-time monitor is not necessarily paired with a test plan. Rather, it is installed permanently and kept in operation during deployment.

The term run-time monitoring is used here in a somewhat broader sense than might be the case in other contexts. A narrow definition would be that a run-time monitor detects anomalous, unsafe, or otherwise problematic behavior and triggers a reaction to mitigate the problem. As a simple example, if the distance to a leading vehicle is too small, a runtime monitor might trigger a command to slow down to increase to a safe following distance. That is within the scope of UL 4600, but is somewhat narrow.

The broader meaning of run-time monitoring used by UL 4600 is any monitoring and reaction to a run-time problem, even if that reaction does not affect AV operation in the moment. In particular, run-time monitoring is envisioned as a method for detecting and reporting some types of SPI violations (see clause 16). So a run-time monitor might detect an operational fault, or might detect an immediately dangerous situation. Or, it might simply detect something that is not quite right per the safety case, even if there is no immediate safety concern. This means that run-time monitoring has a dual role of sometimes reacting to danger and sometimes reporting anomalies. It might even do both at the same time.

The safety argument must use results from run-time monitoring to collect evidence that supports the safety argument. Some of this data might be used as testing evidence. Other data might be used to support arguments that surprises discovered during operation will be detected and acted upon via the use of SPIs.

A key safety principle for run-time monitoring is that any anomalous situation detected by a run-time monitor is considered an incident worthy of investigation, regardless of whether a loss event occurs. This ensures that potential issues with the safety case are detected and addressed promptly rather than waiting for loss events (and, eventually, victims) to pile up over time. This concept will be revisited in the discussion of SPIs, in which some nuance is added to the concept of which statistically relevant events detected by a run-time monitor count as incidents.

12.5. Safety case updates

The safety case must be updated in response to any changes to the system. Since the common case for a change to the system is the discovery of a defect via V&V or run-time monitoring, this topic is discussed along with V&V.

The notion of what a "change" might be is expansive. It includes changes not only to the system design, but also changes to underlying support components (e.g., a security patch to a library), updated tools used to build or manage the compiled software image, changes to data sources, changes to safety analysis tools, a change of item configuration, a change to the ODD, and so on. In general, any change that might invalidate any part of the safety case requires a re-evaluation of and, potentially, an update to the safety case. Additionally, the occurrence of any incident requires revisiting the safety case.

This approach sounds like it puts the need to update and re-assess the safety case on a hair trigger, and that perception is correct. However, safety case updates are subject to impact analysis. This means that only the portions of the safety case that are related to the change need be updated. So the effort required to revisit the safety case is only proportional to the amount the change affects the safety case (including potential ripple effects). The approach to assessing safety case updates responsive to impact analysis is covered by clause 17.

12.6. Resources

- Kane et al., A case study in runtime monitoring of an autonomous research vehicle system, 2015.
  https://users.ece.cmu.edu/~koopman/pubs/kane15_monitoring.pdf
- Pegasus project:
  https://en.wikipedia.org/wiki/Pegasus_Project_(investigation)
- SEBok:
    - Verification:
      https://www.sebokwiki.org/wiki/System_Verification
    - Validation: https://www.sebokwiki.org/wiki/System_Validation

13. Tools, COTS, and legacy qualification

Summary: Clause 13 of UL 4600 deals with the qualification of components, tools, libraries, and other things that can affect safety, but were not developed by the AV design team within the scope of the current project.

Creation of AVs will require not only developing computer-based systems to provide autonomous driving features, but also relying upon Non-Developmental Items (NDIs) to support the engineering process, be used as components within the AV, provide lifecycle support, and so on. NDIs can include hardware, software, services, data, and anything else that is relevant to safety that is not the product of engineering efforts included in the current AV's safety case. In other words, NDIs amount to external dependencies for the safety case. Even legacy software reused from previous products might be treated as an NDI if full information and data are not available for inclusion in the safety case.

One way to treat an NDI is to employ the Element Out Of Context (EOOC) concept described in section 5.2.3. Other potential strategies include tool qualification (e.g., via extensive testing), taking credit for proven in use components, testing in combination with mitigation approaches for any defects potentially missed by testing, and so on. The emphasis needs to be mitigating any way in which the use of an NDI might compromise safety.

The challenge is that an NDI must be brought into the scope of a safety case without being able to argue safety directly based on the engineering processes used to create that NDI. A further challenge is that it is difficult to qualify anything for high integrity levels based on testing alone.

Different types of NDIs have different issues that need to be addressed. A common safety case approach is to identify which NDIs are relevant to safety, what hazards might be presented by their use, and what mitigation within the scope of the AV has been performed to manage any potential risk associated with using NDIs.

13.1. Tools

Any tool that is used in creating, managing, testing, deploying, tracking, or otherwise interacting with the AV, its design cycle, analysis of failures, management of machine learning data, its lifecycle, and so on could potentially undermine safety. In more traditional, test-heavy approaches to vehicle validation it might be thought that tools are of lesser importance because vehicle tests themselves demonstrate safety and ensure the effectiveness of hazard mitigation. However, as soon as the safety case includes arguments related to the adequacy of training data, the use of simulations to provide safety case evidence, the efficacy of field engineering feedback in early detection of problems, and so on, tool qualification becomes dramatically more relevant to the safety case. To be sure, tool qualification has always mattered, but it is likely to be a much bigger issue with AVs.

Each tool that might present hazards (even if said to be mitigated) must be identified including not only the tool, but also the version number and how the tool is used in a way that is relevant to the AV's lifecycle. The version number is relevant because a new tool version might contain defects or introduce new hazards, so the safety case must be revisited every time a tool is updated. This commonly means that design teams are quite reluctant to update tools to newer versions. If tooling is provided via Software-as-a-Service (SaaS) it is worthwhile investigating whether it is possible to avoid automatic upgrades to the tooling and its support infrastructure.

The list of types of tooling that could be relevant to safety is significant, including not just compilers and requirements management tools, but also code analysis tools, machine learning toolchains, configuration management tools, software build scripts, simulators, test campaign management tools, and defect tracking tools, among many others. The key question is whether a tool defect can possibly lead to an incident. If it can, even if the mechanism for that happening is thought to be unlikely to activate in practice, that hazard must be identified, mitigated (or accepted), and tracked to resolution.

Tooling hazards that involve potentially introducing a defect into the AV implementation are the most obvious. That might include a compiler bug, a build tool that fails to enable some static analysis flags, a tool that uses the wrong version of a data set, or a machine learning tool that is misconfigured to use validation data as part of its training data.

Another example hazard is automatically executing a test plan incorrectly, which potentially misses executing some tests or omits reporting test failures. That does not inject defects directly, but rather produces faulty evidence of correctness for the safety case. A related example is a static analysis tool that fails to detect violations it is supposed to detect that might indicate source code defects. Yet another example is a defective field engineering feedback data management system that misses or loses SPI violations which the safety case assumes will be accurately reported. To the degree that the safety case claims that tooling will detect and/or report defects, any potential defects in associated tools become safety-relevant.

13.2. Simulations

It seems likely that simulation results will carry a lot of the evidentiary load for AV safety cases. All the potential issues with tooling in general just discussed apply, along with some additional considerations.

As George Box famously said: "all models are wrong, but some are useful" (see: https://en.wikipedia.org/wiki/All_models_are_wrong) At its core, a simulation is about building a set of models and seeing what they do in a variety of circumstances. The answers will not precisely reflect real-world outcomes (they will in some sense be "wrong"), but they can still be useful if they are predictive of real-world outcomes that matter, such as whether vehicle behaviors will be acceptably safe.

For a simulation to provide useful evidence for a safety case, there must first be an accounting of what parts of the item are encompassed by the simulation. As an example, a simulation that assumes perfect perception without actually simulating a perception system might provide evidence for path planning, but not for end-to-end safety. That is perfectly fine – so long as the safety case does not take credit for end-to-end safety based on the evidence for such simulations.

While the need to express the limits to models (hopefully) seems obvious, accounting for limitations in a composite safety case can be complex. For example, a path planning simulation that assumes perfect perception might still find benefit in modeling sensor occlusions to account for portions of the field of view that even a "perfect" sensor cannot see due to the sensed object being hidden behind some other object. Alternately, the sensor model might account for occlusions and somehow communicate that limitation of sensors to the path planning simulation. However, a safety case in which neither the sensor model nor the path planning model account for sensor occlusions would be invalid.

The representativeness of models in terms of fidelity as well as coverage must be characterized. For example, do the simulation models cover all aspects of the ODD? Do they account for the potentially critical aspects of response times and physical nuances such as surface coefficients of friction for estimating panic braking deceleration?

It is not essential that simulation models represent every possible detail. But what is crucial is that safety credit not be taken for aspects of results that simulations do not cover.

To the degree that simulations are used to replace road testing, it is essential to understand what simulations do and do not provide as evidence, and take credit in the safety case accordingly. As an example, a simulation that covers sun, rain, and snow but not freezing rain might provide plenty of evidence, but closed course tests with artificial freezing rain might also need to be conducted with real vehicles to fill in that gap in simulation capabilities.

Tool support for simulations must also be fit for purpose. Tools that manage experimental campaigns, wrangle model data, and report simulation results must all provide accurate outputs.

13.3. COTS and legacy components

Especially when using machine learning-based techniques, the scope of potential use for Commercial Off-The-Shelf (COTS) and legacy components increases dramatically. (For this discussion, we consider open source components to be a special case of COTS, and do not weigh in on long-standing debates as to the relative merits of different software economic models. Some might prefer the term Off-The-Shelf or OTS.) This category includes hardware components such as sensors, compute platforms, software components, software tools, libraries, network protocol stacks, user interfaces, reused code, pre-trained machine learning models, tools, commercial data feeds, cloud computing infrastructure, commercial data networks, and so on.

The assumption for COTS/legacy components is that the design team will not have access to enough information to construct a safety case, and in the general case an EOOC interface is not available. How to work such components into a safety case is highly dependent upon the system, the situation, and how safety critical the use of the component is.

UL 4600 does not provide a silver bullet for the tricky issue of COTS/legacy qualification. However, it does note that this problem must be addressed in the safety case.

An important caution for COTS/legacy components is dealing with version updates that might potentially be hidden from the AV development team. It is important that any component has a version number that is updated if the version of any internal component is updated. A simplistic "form/fit/function" interchangeability is insufficient for component qualification. As an example, if a lidar sensor is qualified as a COTS component, requalification might be needed if newer units have a new firmware image, a different mask version (with different errata) of an internal CPU executing the same code, a different component used in lidar circuitry, or a different internal power supply design, among other potential differences. Any internal change that might cause a different hazard that must be mitigated by the AV needs to be accounted for in the safety case. Such differences also need to be accounted for in configuration management activities.

The use of machine learning technology provides novel types of COTS/legacy component concerns. As an example, an AV design team might use a pre-trained model as a starting point before applying additional training data. That pre-trained model might have been exposed to biased data, or otherwise have deficiencies that are not readily apparent to the team using it as a starting point. Similarly, COTS data sets might have a non-negligible fraction of inaccurate labels that affect training and validation results. All the inputs and tooling that can affect machine learning-based component performance need to be considered when determining the scope of COTS/legacy component hazards in the safety case.

Open source NDIs are commonly used, especially in the area of machine learning. The allure of "free" software can be irresistible. However, any assurances that such software will be of high quality (e.g., via a "many eyes" argument, or an argument that "everyone" uses the tool so it must be good) are based on a number of assumptions that might not apply to specific tools and specific usage situations. Proprietary tools have their own issues, so this is not to say that open source is better or worse. Rather, any tool must be critically examined to determine suitability for its potential safety implications regardless of provenance.

13.4. Resources

- Wikipedia, Commercial off-the-shelf:
  https://en.wikipedia.org/wiki/Commercial_off-the-shelf

14. Lifecycle concerns

Summary: Clause 14 of UL 4600 deals with mitigating risks associated with lifecycle activities, ranging from initial design to retirement.

While hazards and risks are often broken down by technology or function, a different way of slicing them up is by their role in the lifecycle. This clause of UL 4600 surveys the different types of hazards and risks that can be associated with various lifecycle phases for an AV, but might otherwise be overlooked when taking an approach purely based on design and functionality.

As with most other clauses, the emphasis is on ensuring that the safety case addresses potential safety issues so that they are identified and mitigated. An additional consideration is that since lifecycle phases can extend for many years, there needs to be a periodic mechanism to trigger revisiting the status of risk mitigation to be sure that effective mitigations are still actively being pursued.

14.1. Requirements and design validation

Risks associated with data collection and testing need to be addressed by the safety case. This includes risks presented by testing of all sorts, acknowledging that software defects uncovered during testing might result in unsafe behavior of components or an AV test platform. The presence of a safety operator might mitigate some hazards, but special attention must be paid to the possibility that a malfunction not only causes a dangerous behavior, but also undermines the ability of testing personnel to control that dangerous behavior.

There are two areas of special concern. One is fault injection testing, in which a malfunction is intentionally induced to test the ability of the AV to mitigate any potentially dangerous behavior. As an example, at some point validating the response to a tire blow-out requires actually causing a catastrophic tire failure. If the mitigation fails the test, that leaves anything in the proximity of the AV exposed to unmitigated risk, including potentially being hit by an out-of-control AV. Performing vehicle-level fault injection tests on closed courses instead of public roads is one way to mitigate this type of risk.

The second area of special concern is public road testing, which by its very definition involves exposing public road users to potentially defective AV behaviors. Finding out if behavior is defective is the point of the testing, so any testing failure carries with it inherent risk to the public. As an example of a specific hazard, consider testing on a two-lane public road. If an AV test platform has a defect that causes a sudden, dramatic turn across the centerline into oncoming traffic, the tester might not have enough reaction time (or arm strength) to prevent a crash with another public road user.

Specifics on mitigating public road testing risk are beyond the scope of UL 4600, but readers are recommended to consult SAE J3018 for road testing safety operator practices. Close consideration should also be given to ensuring that any safety operator controls will be effective for mitigating malfunctions given inevitable limits on safety operator reaction capabilities. Other safety practices such as instituting a robust Safety Management System (SMS) are also essential for public road testing safety.

14.2. Build, design release, and manufacturing

There are a number of hazards that can manifest in the handoff from design to manufacturing, many of which have to do with configuration management. It is essential that the hardware and software configuration used by the manufacturing facility is correct, consistent, and uncorrupted. For example, all hardware, software, and components should be specified and checked for version compatibility as part of a formal design release process.

A methodical build process must be created to ensure that the configuration handed off to manufacturing is complete and consistent. Typically this is highly automated to reduce the chance for error and ensure the integrity of each component as well as the entire build package. It is likely that multiple build packages will need to co-exist in a deployed fleet to support multiple different vehicle configurations, so ensuring that each vehicle gets the correct version is essential.

There should also be an automated way to have any manufactured instance report its own configuration as part of an auditing process to check that the handoff was performed properly, and that the configuration of each vehicle is correct and consistent with installed components. This includes both initial design handoffs and updates to ensure that when one component version is updated (for example, a sensor repair part is a newer version), all other components in the system are updated for compatibility (for example, a software driver in the AV's main control computer is installed to properly interact with that new sensor version).

Each manufactured instance of the item must be assembled in a valid configuration, inspected, and tested to ensure that any hazards related to manufacturing faults are mitigated. This will include the use of only approved components, quality control on manufacturing procedures, factory sensor calibration, and so on.

Configuration management is a central concern over the entire lifecycle, but is especially critical here. That includes ensuring not only that the configuration of each vehicle is acceptable for safety, but also that it is possible for the engineering team to determine and analyze what configuration is installed in any vehicle that might be involved in an incident.

Configuration information must not just track the current software and data image in the vehicle, but also be able to trace back to source code, data files, machine learning training data set, and other engineering work products to analyze what issues might have contributed to an incident. While this should sound like a straightforward industry practice, it can be complicated by the use of dynamic data feed services and other off-vehicle item scope. For example, if an anomalous real-time map data feed causes a vehicle malfunction, investigating that issue might require establishing the configuration of a cloud-based data service at the time of the incident.

14.3. Supply chain

Supply chains present unique opportunities to introduce hazards that are easy to overlook during the design process. Third party components might be changed by vendors without notification to the design team. A supplier might suffer quality fade, in which corners are cut on processes and sub-component quality to reduce costs while still finding a way to meet any incoming acceptance tests at the AV manufacturer.

Lifecycle maintenance is subject to supply chain hazards such as the use of unapproved, substandard consumables. Counterfeit repair parts and inferior-grade consumables pose additional challenges, whether used knowingly or inadvertently in an effort to reduce repair and operational costs.

Malicious supply chain attacks might include compromised tool chains that impair the safety of software builds, surreptitious insertion of defects into COTS software components, insertion of unapproved circuitry into integrated circuits, or other ways of undermining safety in ways that are not ordinarily apparent to design and manufacturing teams.

14.4. Field modifications and updates

Field modifications and updates can bring their own hazards. As a practical matter, each AV will need to be continually updated over its operational life to ensure safety. This will happen in response to field engineering feedback via SPIs and in response to changing operational conditions in the real world.

This area adds change control into the mix with configuration management. Changes need to be evaluated for their impact on the safety case as well as AV operation. That will include issue tracking, resolution of problems, and re-validating the system as well as the safety case. To the degree that there are different configurations in operation, each change will need to be validated for all configurations it applies to, and potentially restricted from use on configurations that have not been so validated.

The update mechanism (which might be over-the-air) will need to ensure that any updates installed change the system to a new approved configuration rather than leaving it in a partially updated state that does not match any validated configuration. To the degree that a repair part is installed, that repair part might need to be accompanied by various updates to the repair part and/or the rest of the system to achieve a valid overall configuration.

Things other than the AV hardware and software might be modified or updated. Updates might be made to tools used for maintenance, map database formats, encoded behavior rules (e.g., due to a traffic law change), inspection procedures, maintenance manuals, and test plans. Any update requires revisiting the safety case to ensure that an impact analysis provokes safety case updates as required.

The ability to push out software updates in response to newly discovered problems promises to dramatically reduce the cost associated with recalls to correct design defects. However, it comes with an obligation to be meticulous in ensuring that potential safety issues are mitigated before doing the update.

14.5. Operation

While the safety of performing the driving task itself is covered elsewhere (e.g. clause 7 on interactions and clause 8 on autonomy), there are safety considerations that might otherwise be overlooked included in the lifecycle clause.

As previously discussed, public road testing will be required for the foreseeable future, even after widescale AV deployment. There will always be a software update or ODD expansion that needs to be tested before it is ready for production release. That testing must be performed safely.

A number of other specific considerations are called out to ensure that they do not slip through the cracks. A sampling of them gives an idea of how broad the less obvious aspects of operational safety are:

- **Vehicle and cargo hazards:** safe responses to a vehicle propulsion battery fire, cargo fire, dangerous cabin temperature, unsecured cargo
- **Interactions with other vehicles:** excessive force on unsecured passengers/cargo from a maneuver to avoid collision, impalement by cargo projecting from a leading vehicle
- **Non-operational hazards:** operating a combustion engine in an enclosed space (carbon monoxide poisoning risk) while recharging a hybrid battery, fire while battery charging
- **Stopping motion in an unsafe environment:** stopping on the proverbial railroad tracks, stopping with overheated equipment in contact with a pile of dead tree leaves that poses a fire risk
- **Unauthorized operation:** automated bomb delivery, carjacking, operation in no-vehicle areas, stopping vehicles to intentionally disrupt traffic by blocking streets

As with any other prompt element, the requirement is not to prevent things that cannot reasonably be prevented. Rather, the requirement is to consider operational risks and document mitigation or acceptance within the safety case.

14.6. Retirement and disposal

For an AV to be safe it will almost certainly need to have redundant equipment as well as continual software updates in response to field engineering feedback. But what happens when redundancy is expended or the stream of updates stops? Or when tools, operating systems, cloud infrastructure, or other things essential to keeping the AV in operation become obsolete?

Many AV components will lose their reliability with age. Consumables and wear items such as wiper blades, batteries, protective coverings, and mechanical bearings will need to be replaced or overhauled at some point. If such maintenance is not carried out in a timely manner, the AV will need to restrict itself to degraded operating modes or ensure that its autonomous modes are not activated. If qualified repair parts are not available, that will force a similar outcome. If safety-related maintenance is deferred past a certain point, the AV might need to take itself out of service entirely until the maintenance can be provided.

An analogous situation applies to tools and data. If software updates are no longer available, at some point the AV's safety capabilities will degrade to the point that it should no longer be operated. This includes not only a lack of AV software updates, but also any suspension of monitoring of field engineering feedback to detect emerging problems with the safety case.

More indirectly, if any data feed, service, or other external dependency becomes non-operational, the AV might need to be taken out of service. As an example, if the HD map vendor goes out of business, the AV might become unsafe to operate due to stale maps.

Some conditions might require forced operational suspension for a particular vehicle. A propulsion battery that has accumulated too many charging cycles might need to be taken out of service to mitigate potential battery fire risk until a battery replacement can be performed. Exposure to storm flooding might require permanent retirement due to unacceptable risk from corrosion of connectors and other components.

There will be an economic incentive for an AV owner (for example, someone who has purchased the AV at a deeply discounted scrap price) to attempt workarounds to return the AV to service. Or they might take components that have reached their end of service life and attempt to pass them off as suitable repair parts with a fraudulently stated remaining service life for operational vehicles. How such situations might be handled should be considered in the safety case.

14.7. Resources

- DOT/FAA/TC-15/33 Obsolescence and Life Cycle Management for Avionics
- ISO 28000:2022 Security and resilience — Security management systems — Requirements
- SEBoK: Logistics: https://www.sebokwiki.org/wiki/Logistics

15. Maintenance

Summary: Clause 15 of UL 4600 deals with maintenance and inspection. Activities that are relevant to safety must be defined, performed satisfactorily, and tracked for effectiveness.

Safety-related maintenance of AVs is likely to be substantially more complex than for conventional vehicles. For example, sophisticated sensors will need regular cleaning and calibration. It is also likely to be more critical to do maintenance on time and with high quality due to the lack of a human driver to detect incipient problems and intervene when things go wrong.

As a simple example, if a conventional vehicle driver suffers degraded visibility due to running out of windshield wiper fluid, we blame the driver for failing to refill the reservoir. But how do we ensure that an AV does not run out of sensor lens cleaning fluid while driving? And how do we deal with that sensor being gouged by a rock kicked up by another vehicle, being twisted out of alignment via a strike from a low-hanging tree branch, or being covered by a wind-blown opaque plastic bag that obscures its field of view? Surely such situations can be handled, but it will take more planning than counting on a human driver to notice, compensate for the problem, and know to schedule a service appointment if required.

Any aspects of maintenance and inspections that affect safety must be included in the safety case. Put another way, if any part of the safety case takes credit for maintenance and inspection, there must be an argument that those actions will be done consistently and correctly. While it is all well and good to posit that someone will notice and correct an AV equipment problem, it is important to ensure that this actually gets done in a timely and thorough way.

15.1. Maintenance and inspection requirements

The safety case must include any maintenance and inspection requirement that is relied upon when arguing safety. There must also be a fault model associated with those activities that includes, at a minimum, procedures being deferred, omitted, performed incorrectly, or performed too often ("too often" is potentially relevant if inspection procedures induce component wear).

Maintenance and inspections are likely to be required during a number of lifecycle stages. Some inspections need to be done at the factory during the initial commissioning of each vehicle, then selectively revisited after a major repair. Other inspections need to be done periodically while in service to detect worn components, check for satisfactory calibration, and exercise components that cannot be self-tested while in normal service. Periodic maintenance will be needed to replace wear items, clean, and replenish consumables.

Each maintenance and inspection activity must have a defined procedure that encompasses everything the safety case takes credit for. The inspections might be a mixture of manual, semi-automated, and fully automated procedures.

There also needs to be some validation of procedures. They need to be written in a way that ensures they will be performed as intended in real-world conditions by suitably qualified maintenance personnel – and that the personnel are in fact suitably qualified. Sometimes maintenance might be done by a vehicle owner, but specialized procedures are likely to involve a technician certification process. There should also be a quality process in place to ensure that maintenance outcomes are as intended in the operational fleet.

15.2. Prompting maintenance and inspection activities

It is essential to establish when and/or how often various maintenance and inspection activities must be performed. The safety case must argue that each such activity is performed often enough to result in an acceptably maintained system. It's one thing to plan on maintenance, but it is much more involved to set up a system that makes sure that maintenance (and inspection) gets done in a timely way in real-world usage.

Inspections are likely to be periodic based on one or both of time and vehicle usage. This might include calendar time (once per week/month/year), operational time (hours of usage), number of equipment operational cycles, and/or distance traveled. These triggers account for a variety of failure causes that span wearout, consumable consumption, component aging, and random failures.

Beyond preventive maintenance, managing redundancy as part of ensuring safety requires knowing when a component has failed so that it can be repaired or replaced. Prompting such maintenance requires a system to detect component failures, either by noting their failure during operation, and/or by scheduling periodic inspections to check for latent faults. Inspections might be required to ensure that components are healthy if failures are difficult to detect in other ways.

Sensors that are intended to detect faulty equipment are especially difficult to test in service, and will need to be subjected to periodic inspections to ensure that they continue to be able to detect faults. As a simplistic example: How do you determine that an automatically activated wiper system is operational if the vehicle has not encountered rain in months due to operations during the dry season? Perhaps you spray water on it in a garage as an inspection procedure to see if it activates.

There will need to be a way to track the triggers for various types of inspections and preventive maintenance, as well as record when a detected failure has triggered a need for corrective maintenance. Some of the triggers might be conditional (for example, inspect wiper operation if the wipers have not been activated due to operational situations for a month). There might be a tradeoff between complexity and cost (e.g., perhaps inspect wiper operation monthly regardless of whether it has rained to streamline and simplify procedures).

The safety case must argue that even though it is likely to be complicated to administer, the required maintenance and inspection regime will be carried out in a timely, accurate fashion for every vehicle.

15.3. Maintenance and inspection faults

Maintenance and inspection can go wrong, so the safety case must argue that potential faults with those procedures will be sufficiently rare.

Procedures might be omitted in whole or in part because of a scheduling error. Or perhaps they will be missed due to an inattentive vehicle owner, a short-handed maintenance crew under pressure to return vehicles to revenue service, or even illicit cost-cutting.

Procedures might be incorrectly completed due to carelessness, honest mistakes, or the use of unapproved repair materials. For example, a supply chain compromise could insert substandard materials into official parts supplies. A repair parts shortage could motivate maintenance personnel to use refurbished or aged components that are still functional (but insufficiently reliable) to meet system uptime quotas – or simply to improve profitability.

Procedures might be incorrectly completed, partially skipped, or completed using unapproved shortcuts to save time, meet productivity goals, or simply permit workers to avoid undesired overtime work. Those taking shortcuts might not even be aware of dependencies in the safety case that require a lengthier process and feel that they are cleverly saving time and money by avoiding work they believe is not useful.

If procedures have not been performed correctly, a fraudulent statement of procedure completion might still be reported to tracking systems. For procedures that can affect safety, there will need to be some sort of check and balance to ensure the timely, effective completion of safety-related maintenance and inspection procedures by qualified personnel.

15.4. Non-operational safety

A potentially overlooked area of maintenance and inspection relates to non-operational equipment failures. These can be generally broken down into hazards related to an inactive vehicle and equipment degradation.

15.4.1. Inactive vehicle

Even though an AV has completed a transportation mission and is currently inactive, there are still hazards to be considered. Those hazards include:

- **Occupant egress.** If a passenger is somehow trapped in a vehicle when it is taken out of service, how do they get out? This is especially of concern in environments that can result in very hot vehicle interiors.
- **Recharge/refuel.** The vehicle must not exhibit dangerous behaviors while in a recharge/refuel state. Potential hazards include dangerous emissions in an enclosed garage space (hydrogen from lead/acid battery recharging, carbon monoxide from combustion engine used to recharge), and insufficient mitigation of potential battery fires. A related hazard would be returning to service without ensuring that any charging/fueling connection has been properly disengaged. (The astute reader will realize this type of hazard has been discussed previously as a lifecycle concern. Recharge/refuel might be done mid-mission, or might be done as an overnight maintenance activity, so it is covered in both places to make sure it is not missed. It is fine for one argument about refuel/recharge safety to support multiple prompt elements in UL 4600.)
- **Non-operational movement.** The vehicle or its components will need to be operated and moved in special ways in support of maintenance, inspection, and vehicle transport. The AV needs to react in quite different ways in these conditions compared to normal operation. Incorrect operation will pose a hazard to maintenance and transportation personnel, and must be considered in the safety case. At the very minimum, it must be ensured that "disable" mechanisms used by such personnel have high safety integrity. Additionally, maintenance and inspection procedures must pair with safety mechanisms to ensure that unexpected movement during procedures does not pose a physical hazard to technicians. (Again, a topic discussed before but covered from a different angle.)

15.4.2. Equipment degradation

Some components might fail if unused, especially for long periods of time. For example, a system that is designed to be operated weekly might lose its lubrication film over a period of months and be subject to corrosion that would not otherwise occur. A battery that is intended to be recharged every drive cycle might lose charge over a long period of inactivity. And so on.

Additionally, non-operational equipment might be subjected to conditions that compromise its safety in ways that it cannot readily detect by the vehicle itself because it is not actually active at the time. Misfortunes such as dirt accumulation, floodwater exposure, overheating, excessive cold, physical sensor damage, being hit by another vehicle, vibration damage from air transportation, infestation by insects, and cable integrity compromise by chewing vermin can all occur.

Some combination of inspection and maintenance is likely to be required after any substantial period of non-operational status. That includes preparation for deployment after transport from a manufacturing facility, significant periods of disuse due to low fleet demand, being parked at an airport parking lot for an extended time, and spending a long time in inactive storage while awaiting repair.

A related type of "degradation" is a lack of updates. Vehicle safety updates might not be processed if the vehicle is in an extended inoperative state. These might include missed over-the-air software updates as well as performing remedies for equipment recalls. (For example, the vehicle is at a repair facility and misses the shift in which all other vehicles have had maintenance performed to implement a recall remedy.) Part of recovery from inactivity should be a check to ensure that all updates and corrective maintenance that might have been missed have been performed.

15.5. Resources

- SEBoK: System Reliability, Availability, and Maintainability: https://www.sebokwiki.org/wiki/System_Reliability,_Availability,_and_Maintainability
- Wikipedia, Maintenance: https://en.wikipedia.org/wiki/Maintenance_(technical)

16. Safety Performance Indicators

Summary: Clause 16 of UL 4600 deals with metrics and, in particular, Safety Performance Indicators (SPIs). SPIs are used to monitor and drive improvement to the AV and its safety case across the lifecycle.

The heart of the UL 4600 approach to field engineering feedback is the Safety Performance Indicator (SPI, pronounced "S-P-I") mechanism. An SPI is defined as "a metric used to quantify safety performance." (See definition 4.2.39 in UL 4600 version 2.) While some obvious metrics include how often a crash might occur, other metrics are much more indirect, such as how often a "surprise" object is encountered that confuses the AV perception system, or how often a safety case assumption might be violated.

The idea for feedback is to instrument the AV and safety-related processes (including the vehicle, infrastructure support, and lifecycle support) to measure when things are happening that undermine the safety of the system. The term "safety performance indicator" is used in the aviation domain to refer to operational safety metrics. UL 4600 expands that notion to include safety metrics for the entire lifecycle, from design through the disposal of the item.

16.1. Threshold values

To be useful, an SPI must not only define a metric calculation approach, but also have a defined threshold value. This relates to the general "how safe is safe enough" question.

Suppose that an AV mistakenly runs through a red traffic signal instead of stopping, and that happens once every 2 million miles. Is that safe enough? The answer is "it depends" because we need to know how that rate fits into the risk budget, and perhaps how it might compare with human drivers making the same mistake. Missing a red traffic signal is dangerous to be sure, but not every missed traffic signal results in a crash. So the question is how often such a miss might be tolerated while reaching an acceptable system-level safety outcome. While an engineering goal of "never" is desirable, it is quite likely that real-world outcomes of "extremely seldom" might be good enough.

Similarly, an improvement on a metric without knowing the threshold does not necessarily mean much. Using that same hypothetical example, improving to one red light violation per 4 million miles is twice as good. But it might still be terrible if the necessary threshold for acceptable safety is once every 20 million miles. Or it might be a nice but only slight improvement to safety if the threshold was once every 100,000 miles and other factors dominate the risk profile.

A threshold value typically involves some frequency, rate, probability, or other ratio of problematic situations compared to acceptable situations. This might be events or time spent in dangerous conditions tallied per operational hour, per vehicle kilometer traveled, per scenario handled, per object encounter, per component malfunction, and so on. It is unreasonable to expect absolute perfection in any real-world system, so the threshold even for highly critical components will be some non-zero ratio, albeit often an extremely small ratio of near-perfection.

UL 4600 does not set specific permissible dangerous event ratios for safety, leaving that up to the creators of the safety case. However, other safety standards commonly use numbers in the one per million to one per billion per hour range for many types of failures that escape past mitigations to cause incidents. In some cases it might be convenient to use an SPI threshold of zero for catastrophic events rather than expend engineering effort computing an extremely small allowance for something that, in practice, should never happen.

16.2. Dangerous behaviors

SPIs should be defined for system behaviors that are causally connected with vehicle risk. Examples of item-level SPIs that directly measure undesirable outcomes include fatalities, other crashes, violations of traffic rules (e.g., running a stop sign), and near hits (safety margin violations that get lucky and do not result in a loss event). Generally these correspond to so-called "lagging" indicators that measure safety outcomes during operation.

Some malfunctions, failures, or other undesirable situations might not present an obvious risk to vehicle occupants and other road users in the moment. SPIs for those might be more of a "leading" indicator that can be predictive of potential future loss events. Any SPI that is not associated with an overtly dangerous situation should be able to serve as a leading indicator for safety. As a simple example, a sensor that malfunctions much more often than it should is likely to eventually fail at just the wrong time, potentially contributing to a crash. But earlier failures will just be a leading indicator of a potential problem.

To the degree the safety case has argued that certain hazards and risks will be mitigated, a failure to actually mitigate them is a material failure of the safety case. As an example, if the safety case argues that two cameras will not fail at the same time due to a lack of common cause failure modes, then having those two cameras fail together undermines the credibility of the safety case – even if other sensors are still available that permit loss-free operation of the AV on observed multi-camera-failure missions. Similarly, if individual cameras fail more often in use than expected, that will likely increase the probability of random chance failures in two cameras that happen to occur on the same mission even if not occurring at exactly the same time.

Other performance and quality failures can increase operational risk even if they do not lead to an incident in the short term. That might include perception failure rates (elevated false positive or false negative detection rates), software component execution faults, real time scheduling deadline misses, and failed "sanity" checks on data values that should never present out of bounds values. A robustly designed system can often tolerate many such faults. Nonetheless, any such event can cast doubt upon the integrity of the system, and in particular the soundness of the safety case.

A robustly safe system will be designed with multiple levels of safety mechanisms and failsafes to reduce the probability of incidents and loss events. While any particular failure might be mitigated by such a defense-in-depth approach, that does not mean that the activation of a mitigation mechanism can be ignored. The rate at which failsafes are activated should be measured with SPIs to give advance notice that safety-related failures are in fact occurring – before the day arrives in which all the protection layers fail and there is a loss event.

SPI-relevant data sources are not limited to strictly technical failures. Process failures can be associated with SPIs. Most importantly, the safety culture of the organization should be subject to SPIs, including measuring process adherence (design processes as well as maintenance/inspection processes), and ensuring personnel performing processes have appropriate skills. Any field defect or incident should be traced back to see if deviations from processes played a role as an indirect, root cause analysis SPI metric source.

16.3. Surprises

Another use for SPIs is detecting things that are unexpected or thought to be impossible, which are generically termed "surprises." In keeping with the notion of an SPI threshold, the arrival of any particular surprise might not be problematic if it happens infrequently enough. The key is whether surprises are happening so often that the soundness of the safety case is called into question.

One type of surprise is encountering situations in which the ODD used for designing the system does not match the actual intended operational environment. This might result in a surprise ODD departure, such as encountering mapping errors, objects that do not follow behavioral prediction probability envelopes, anomalous driving behaviors by other road users, and potentially even faulty vehicle behaviors from the AV's own vehicle.

At its heart, every surprise is an invalid assumption that is either explicitly or implicitly encoded into the safety case. Some surprises violate an assumption that a certain problematic condition will be rarer than it really is. Other surprises violate an assumption that the environment is understood well enough that there will be no unforeseen environmental conditions (i.e., there are essentially no remaining "unknown unknowns" that will be encountered in operation). Still other surprises might arise from unexpected behavioral interactions, such as a risk mitigation behavior (e.g., forceful braking to avoid a collision) precipitating other hazardous situations (e.g., potentially being hit from behind while forcefully braking).

16.4. SPI collection and feedback

SPI data will need to be provided from a wide variety of sources. The most obvious source will be onboard vehicle instrumentation. This means that the AV architecture will need to include detection and logging mechanisms for SPI-relevant events. Other SPI data sources might include design process quality monitoring, a Safety Management System (SMS) for operational safety, road testing results, and so on.

Logged data will need to be transmitted to an engineering center for processing and analysis. Data integrity will need to be supported throughout the process to ensure not only data accuracy, but also metadata correctness (e.g., which vehicle, under what operational conditions, for which SPI, for which safety case version, is associated with a particular SPI datum).

The collection of SPIs will need to be periodically monitored against defined threshold values. That monitoring should identify outright SPI violations that invalidate the safety case. However, monitoring should also look for abnormal trends as an input to statistical modeling, and as a source of data for optimizing maintenance periodicity as well as optimizing the safety case as experience is gained.

Non-SPI data and analysis should be used as a check on the effectiveness of the set of SPIs. For example, a loss event might occur without any warning sign that there was a gap or defect in the safety case. An analysis of any such event might identify new SPIs to be added to instrument parts of the safety case that are potentially related to the crash. Those SPIs could then monitor potentially related areas of the safety case during testing and further operation to ensure that any corrective actions taken in fact reduced the risk of another similar loss event. In practice, it might be that a safety case starts with somewhat sparse SPI instrumentation, then has more SPI instrumentation added based on experience from root cause analysis of anomalous testing and operational events.

While some proposed SPIs might be statistically tied to safety rather than causally linked, it is better to use SPIs that have a causal connection to safety. As an example, hard braking for a human-driven car might be statistically predictive of an increased risk of collision. But the hard braking itself might not have a causal connection to that risk. Sure, hard braking might get a human-driven vehicle hit from behind, but is also likely to be a symptom of aggressive driving, inattention, or operation in an especially risky geographic area full of other erratic road users. Selecting hard braking as an SPI might not correlate with safety for an AV. Consider an AV that has a maximum braking force limit to prevent hard-braking SPI violations,

Safety Performance Indicators

resulting in it suffering collisions whenever a car in front of it brakes aggressively or a pedestrian darts out into the road. While such a metric might superficially be said to be an SPI, if too easily gamed it will lose its predictive power for safety. Such metrics should be avoided.

16.5. A proposal for tying SPIs to the safety case

SPIs are only useful in practice if they have the property that an SPI threshold violation sheds insight into a potential problem with safety that can be fixed. This is the entire point of using SPIs to drive field engineering feedback.

Similarly, a lack of SPI violations during testing and deployment builds confidence in the soundness of the safety case over time. For that to be true, the SPIs should be tied directly to the safety case in some way, so that a violation of the SPI threshold of necessity demonstrates that there must be a problem with some part of the safety case.

Version 3 of UL 4600 (undergoing a consensus process at the time this is written) proposes a more specific formulation of an SPI as a Highly Recommended prompt element. That formulation is:

> *An SPI is a metric supported by evidence that uses a threshold comparison to condition a claim in a safety case.*

This definition can be broken down term by term:

SPI (Safety Performance Indicator): A {metric, threshold} pair that measures some aspect of safety relevant to an autonomous vehicle according to the stated definition.

Metric: A value, typically related to one or more of product performance, design quality, process quality, or adherence to operational procedures. Often metrics are related to time (e.g., incidents per million km, maintenance mistakes per thousand repairs) but can also be related to particular release versions (e.g., significant defects per thousand lines of code; unit test coverage; peer review effectiveness).

Evidence: The metric values are derived from data measurement sources.

Threshold: A metric on its own is not an SPI because context within the safety case matters. For example, the number of false negative detections on a sensor is not an SPI because it misses the part about how good the metric has to be to provide acceptable safety when fused with other sensor data in a particular vehicle's operational context. ("We have 1% false negatives on camera #1. Is that good enough? Well, it depends...") There is no inherent limit to the complexity of the threshold determination if such complexity is justified. But in the end, the answer is some sort of comparison between the metric and the threshold that results in true or false. An SPI threshold comparison being false means that an SPI violation has occurred.

Condition a claim: Each SPI is associated with a claim in a safety case. If the SPI threshold comparison is true, the claim is supported by the SPI. If the SPI threshold comparison is false, then the associated claim has been falsified. SPIs based on time series data could be true for a long time before encountering a situation that makes them go false, so this is a time- and state-dependent outcome in many cases. However, once an SPI threshold violation has occurred, the claim has been falsified and that version of the safety case has been rendered permanently unsound. This can only be remedied by changing the safety case (and, potentially, changing the AV, its operational environment restrictions, or some other characteristic that will be reflected in the safety case).

Safety case: Per ANSI/UL 4600 a safety case is "a structured argument, supported by a body of evidence, that provides a compelling, comprehensible and valid case that a system is safe for a given application in a given environment." In the context of that standard, anything that is related to safety is in the safety case. If it is not in or referred to by the safety case, it is by definition not related to safety.

A key idea with this type of SPI is that it is inherently coupled to the safety case by being tied directly to a specific claim. Each SPI has the purpose of monitoring one very specific claim to see if that claim has been falsified. That way, any time an SPI violates its threshold, it is a sure thing that a problem has been found with the safety case.

An additional aspect of this formulation is that it makes it more clear that an SPI can measure anything relevant to safety, even if the connection to any specific loss event might seem tenuous. If one of the claims of safety rests on following a specific software coding style standard, then violations of that coding style standard trigger an SPI violation – even if nobody would be able to say what the exact link might be between that violation and a potential vehicle crash. If the safety case relies on a claim to build its case for safety, then an SPI monitoring the truth of that claim is by definition also relevant to safety.

16.6. Resources

- Koopman, P., How Safe Is Safe Enough? Measuring and Predicting Autonomous Vehicle Safety, September 2022.
    - Chapter 8 covers a way to link SPIs to safety case claims in detail.
    - https://safeautonomy.blogspot.com/2022/09/book-how-safe-is-safe-enough-measuring.html
- Koopman, Safety Performance Indicators and Continuous Improvement Feedback (SEAMS Keynote talk), 2022
    - Slides: https://users.ece.cmu.edu/~koopman/lectures/L127-2022-04-SEAMS-Feedback-SPIs.pdf
    - YouTube: https://youtu.be/mRXotHN0Z6I

- Archive.org: https://archive.org/details/1127-safety-performance-indicators-and-continuous-improvement-feedback_202205

17. Assessment

Summary: Clause 17 of UL 4600 covers the assessment of the safety case. There are two different phases of assessment: self-assessment and independent assessment. Ultimately the outcome of the assessment processes is, if appropriate, a statement of conformance to the standard.

The overall purpose of UL 4600 is to provide a structured, granular way for an assessor to evaluate an AV safety case. The other clauses cover how the safety case should be constructed as well as topics that ought to be addressed by the safety case. This clause talks about the assessment processes that are used to check that a particular safety case in fact conforms to all of UL 4600's requirements.

During the consensus process for the first version of UL 4600, there was spirited discussion about the relative capabilities, limitations, and incentives of different parties in determining whether a safety case was acceptable. On the one hand, the design team knows the most about the AV they are building since it is their own work. On the other hand, an independent assessor is less likely to have confirmation bias, and might have more diverse experience to draw upon while performing an assessment. Both design teams and independent assessors were thought to be prone to somewhat different types of subjectivity and variability in assessment outcomes for a variety of reasons.

Rather than pick only one strategy for assessment, both self-assessment and independent assessment are required so as to harness the strengths of each approach. They are arranged in a way that is intended to minimize redundant work and improve consistent outcomes.

There are three main assessment activities:
- Conformance planning
- Self-assessment
- Independent assessment

Ultimately the result of these activities should be a safety report indicating that the safety case for a specific AV release is complete, valid, and sound. That safety case will need to be updated and re-assessed throughout the AV's lifecycle to maintain conformance.

17.1. Conformance planning

There must be a conformance plan to ensure that conformance assessment is done in a structured, methodical way. The core of the conformance plan is ensuring that the safety case addresses all the prompt elements in UL 4600, subject to deviations permitted by UL 4600 clause 4. Recommended prompt elements need not be addressed, but all other prompt elements must be addressed by the safety case one way or another. When discussing traceability below we omit this Recommended prompt element exception for brevity.

Lists required by prompt elements must be complete (e.g., "identify" in a prompt element means to identify all relevant system components). Also, there must be full traceability to resolution, such as ensuring that all identified hazards are documented in the safety case as having been accepted or mitigated.

The result of creating the safety case and performing conformance assessments is a conformance package that contains: the conformance assessment plan, the complete safety case, traceability of the safety case to each prompt element in UL 4600, and the assessment results. In particular, every prompt element in UL 4600 must be traced to a specific argument within the safety case (subject to permitted deviations).

The conformance package must contain all the materials needed to establish conformance. For a conformant safety case if one asks, for example, "what is the data transmission fault model for this safety case?" the answer is not found by searching through the safety case to see if it can be found. Rather, the answer should be going to an index of UL 4600 prompt elements associated with the safety case to locate a traceability feature for UL 4600 prompt element "6.2.6.1.b. data transmission faults." That should reference the specific sub-argument within the safety case that describes the data transmission fault model. Moreover, there should also be traceability from other parts of the safety case related to data transmission to that same sub-argument. An assessor can then check to see if the identified sub-argument indeed addresses all the relevant prompt elements of UL 4600, and that any relevant hazard analysis in the safety case fully accounts for the stated fault model.

In other words, checking for conformance should be an exercise in following established traceability links in the safety case and evaluating whether the arguments and evidence make technical sense. Conformance checking should not devolve into an unstructured treasure hunt to see what support for safety might be found.

Safety reports become a part of the system history and are retained with the safety case as part of the conformance package for future assessments.

17.2. Self-assessment

A key concept for UL 4600 is that the primary burden of assessment and final responsibility for safety is on the design team, not the independent assessors. This is viable because the design team indeed knows their own system the best. Self-assessment can be completely integrated into the design process, form a separate parallel process, or any other approach that the design team feels is best. There is no independence requirement for the self-assessment process.

In practice, safety self-assessors are likely to be involved at various times in creating the safety case, analyzing evidence to support the safety case, establishing traceability across the pieces of the safety case, updating the safety case responsive to SPI data, and polishing the safety case to prepare for independent assessment. This will require that members of the self-assessment team have suitable safety and system engineering skills.

There is no need to wait until the AV design is almost complete to start self-assessment, and indeed waiting is likely to be a poor strategy. The need to create a safety case with supporting evidence is likely to influence architectural choices up-front in the AV design process. A good strategy is to co-evolve the AV design and the safety case. Having the self-assessment team work on the safety case during that co-evolved design process is acceptable because there is no presumption of independence for that team.

The philosophy should be that the final self-assessment round is in fact the team's best effort at a rigorous assessment of the safety case. This means that the self-assessment team should conduct a complete mock independent assessment and resolve any discrepancies as preparation for the real independent assessment.

17.3. Independent assessment

In principle, one might argue that a self-assessment would be all that is required. But such an approach is unlikely to result in safety in any real-world product. Independent checks and balances are an absolute requirement for real-world safety. There are simply too many potential ways a non-independent self-assessment can go wrong, ranging from unwitting confirmation bias to management pressure to deploy per schedule. Thus, UL 4600 also requires an independent assessment.

The independent assessment is done by specialists in safety assessment who have sufficient knowledge in the relevant technology and application domain. Their task is not to create the safety case, nor to find evidence of safety. Rather, their case is to perform a critical review and audit of the safety case provided to them as an output of the self-assessment process. Confirmation bias is still a consideration, and the independent assessment should not simply boil down to an audit. Rather, there needs to additionally be a critical review of the substance of the safety case.

While there is no strict separation between audit and critical review, some activities of independent assessment will be more like a book-keeping audit. Examples of audit-style activities are ensuring that every prompt element in the standard is accounted for in the safety case, and ensuring that every entry in the hazard log refers to a resolution argument in the safety case. Orphaned arguments and dangling references are structural defects in a safety case that should be found by an audit.

Audit tasks can be somewhat mechanical, and to the degree practical, might be automated. In an ideal world, there is an audit tool that automates everything that can reasonably be automated. That tool provides an audit report highlighting discrepancies and places in which the audit tool cannot be sure about conformance. In principle, the audit tool should be able to find an outright discrepancy during an independent assessment, because those should have been caught and fixed as part of the self-assessment process when the same tool was run. (If such a tool is proprietary to an independent assessment organization, they might run the tool and have the design team correct any discrepancies before spending resources on further assessment activities.)

The critical review part of the independent assessment is where judgment and experience come in. In this activity, the independent assessors evaluate the credibility of the safety case. Are the arguments reasonably complete? Are assumptions plausible? Is the logical flow of the argument free of fallacies and general hand-waving? The independent assessors are required to push back on the design team where they find the safety argument to be lacking in technical substance, accuracy, or credibility. To do this, the independent assessors must have sufficient technical competence to understand and critique the safety case.

The term "independent" is to some degree relative. In a specialized technical field there necessarily will have been some contact between team members and assessors, such as previous conversations at professional conferences. And there will always be some manner of financial entanglement. If nothing else, the assessors will need to be paid for their work, and will need repeat business over the long term to remain viable.

Nonetheless, there are clear warning signs of a lack of independence that need to be avoided. Some examples of clear red flags are contingency payment (payment only if/when conformance is approved), management conflict of interest (independent assessor group's supervisor has a shared management reporting structure to a product manager who is responsible for putting the AV into production), and historical entanglement with the design process (the independent assessors provided advice on how to build the AV).

The use of an internal (but independent) team is permissible under UL 4600. Opinions vary as to whether an external, third-party independent assessment team is advantageous, but a high level of independence must nonetheless be argued whether the team is internal or external to the company designing the AV. Using independent assessors who have themselves been accredited by a third party for that role is Highly Recommended.

It is crucial to avoid the negative dynamic of the design team (including self-assessors) creating a lackluster safety case to see what they can get away with in independent assessment. This negative dynamic can escalate to a role reversal in which the independent assessors are blamed for missing a safety defect that later causes a loss event. (A similar dynamic is commonly seen when defective software causes a loss event and blame is placed on testers for having missed the defect, rather than developers for having created the defect in the first place.)

Full responsibility for safety falls squarely on the combination of the design team and self-assessors. The independent assessors form a check and balance that can provide significant value. Nonetheless, good faith effort by independent assessors is sufficient, and does not create a presumption of culpability even if they miss something.

17.4. Continual re-assessment

It is inevitable that the safety case will need to change over time. Those changes will occur because of improvements or changes in functionality, expansion of the AV's ODD, responses to SPI threshold violations, or changes in the outside world that necessitate changing AV functionality.

Changes will trigger a need for various types and scopes of reassessment, which should be done according to a conformance monitoring plan. That plan should provide for triggering self-assessment activities as well as independent assessment activities. There should also be a written plan for escalating non-conformance issues if not resolved in a timely manner, potentially escalating to fleet stand-downs if non-conformance issues are too severe and take too long to resolve.

Every change to the safety case and update to evidence will trigger a need for a renewed self-assessment. That self-assessment should be informed by impact analysis to manage the scope required. As an example, if a new batch of operational data, simulation results, or other engineering analysis data is entered into an evidence repository, all claims about that evidence need to be re-evaluated to see if they are still true. To the degree practical such routine checks might (and probably should) be assisted by automated tooling. Nonetheless, it needs to be done and an exception raised to the self-assessment team if a claim becomes invalid for any reason.

Maintenance of the safety case similarly should consider the impact of any change and re-assess whether potentially affected fragments of the safety case also need to be updated, with any potential ripple effects across the safety case taken into account.

In principle, there needs to be a new self-assessment after every safety case change, including changes to claims, argument, and evidence. However, in practice, it is acceptable to only self-assess the portions of the safety case that seem likely to be affected, as informed by a defined impact analysis process.

In some cases an independent assessment will also be required. An independent assessment should be scheduled periodically, because a series of "small" changes will over time cumulatively induce a potentially "large" net change in the safety case. Additionally, independent assessments should be performed when any single change is "large" according to measures such as a dramatic change to functionality, use of a fundamentally different technology approach, operation in a significantly new environment, modification to high risk/high integrity functionality, and so on. (The characteristics of "small" vs. "large" changes are at the discretion of the design team, but must be documented in the initial safety case before such changes might take place. "Small" vs. "large" decision criteria are subject to critique by independent assessors.) Independent assessment can be informed by impact analysis as well, depending on the nature and scope of the change.

A final task in continual re-assessment is incorporating lessons learned. The design team should update its safety case to add custom prompt elements not included in UL 4600 based on its lessons learned. Assessors should, when possible, use lessons learned to submit generic prompt element suggestions to the UL 4600 standards update process for inclusion in a new version of the standard.

18. Wrap-up

While this book has summarized many of the key ideas of the standard, inevitably there are pieces missing or simplified. There is nothing like reading the standard itself to know what is there. To that end, we discuss how to access the standard free of charge and provide other support and background information.

18.1. Additional information

Further information about UL 4600 can be found on this web page:
https://users.ece.cmu.edu/~koopman/ul4600/index.html

At the time of publication of this book, key resources presented on that web page include:
- A downloadable voting draft copy of version 1 of UL 4600. Important changes have been made since that time. However, the general structure and substantial amounts of content remain similar or identical to the issued standard. This is a convenient way for educators to give students an easy-to-use copy of the standard, as well as for new users of the standard to get a convenient orientation free of charge.
https://users.ece.cmu.edu/~koopman/ul4600/191213_UL4600_VotingVersion.pdf
- Access to the free digital view of the current official version of the standard via the UL standards web site. There is a link to purchase an official copy, but the "digital view" link permits web viewing of the entire standard with no purchase. Free account registration is required.
https://www.shopulstandards.com/ProductDetail.aspx?productid=UL4600
- A video overview of the standard (20 minutes):
YouTube: https://youtu.be/ymGovH_K56U
Archive.org: https://archive.org/details/L109-ul-4600
- A list of current voting STP members:
https://csds.ul.com/Include/ViewRoster.aspx?GroupID=1862&stpNumber=4600
- A frequently asked questions (FAQ) page:
https://safeautonomy.blogspot.com/p/ul-4600-faq.html

More general support material on safety critical system design lessons learned can be found here:
- Compendium of automotive software recalls: https://betterembsw.blogspot.com/p/potentially-deadly-automotive-software.html
- Computer-based system safety essential reading list: https://safeautonomy.blogspot.com/p/safe-autonomy.html
- Koopman, Safe Autonomy Blog: https://safeautonomy.blogspot.com/
- NASA Safety, Quality, Reliability, Maintainability library: https://standards.nasa.gov/safety-quality-reliability-maintainability
- The RISKS Digest: https://catless.ncl.ac.uk/Risks/

18.2. About the author

Prof. Philip Koopman is an internationally recognized expert on Autonomous Vehicle (AV) safety whose work in that area spans over 25 years. He is also actively involved with AV policy and standards as well as more general embedded system design and software quality. His pioneering research work includes software robustness testing and run time monitoring of autonomous systems to identify how they break and how to fix them. He has extensive experience in software safety and software quality across numerous transportation, industrial, and defense application domains including conventional automotive software and hardware systems. He was the principal technical contributor to the UL 4600 standard for autonomous system safety issued in 2020. He is a faculty member of the Carnegie Mellon University ECE department where he teaches software skills for mission-critical systems. In 2018 he was awarded the highly selective IEEE-SSIT Carl Barus Award for outstanding service in the public interest for his work in promoting automotive computer-based system safety. In 2022 he was named to the National Safety Council's Mobility Safety Advisory Group.

Web link: https://users.ece.cmu.edu/~koopman/

www.ingramcontent.com/pod-product-compliance
Lightning Source LLC
Chambersburg PA
CBHW050007230526
45465CB00003BB/1296